Nanotechnology Perceptions
A review of advanced technologies and their impacts
ISSN 1660-6795
E-ISSN 2235-2074

PUBLISHER

The Netherlands Press, Amsterdam

EDITORIAL OFFICE

E-mail: editor@nano-ntp.com

EDITORIAL BOARD

Dr Estefanía Abad, Consorcio ESS Bilbao, Derio, Bizkaia, Spain
Prof. Christopher M.A. Brett, Universidade de Coimbra, Portugal
Dr Marcello G. Cacace, Institute for the Study of Nanostructured Materials, Bologna, Italy
Prof. Che Ting Chan, Hong Kong University of Science and Technology, Hong Kong
Prof. Marc Desmulliez, MicroSystems Engineering Centre (MISEC), Heriot-Watt University, Edinburgh, Scotland
Prof. Kurt E. Geckeler, Gwangju Institute of Science & Technology (GIST), South Korea
Dr Graham Holt, Collegium Basilea (Institute of Advanced Study), Switzerland
Prof. Paata J. Kervalishvili, Georgian Technical University, Tbilisi, Georgia
Prof. Toyoki Kunitake, Faculty of Environmental Engineering, University of Kitakyushu, Japan
Prof. Krishan Lal, National Physical Laboratory, New Delhi, India
Prof. Athanasios G. Mamalis, National Technical University of Athens, Greece
Prof. Lars Montelius, Department of Physics, University of Lund, Sweden
Dr Michael Neumann-Spallart, Dept Inorg. Technol., University of Chemistry & Technology, Prague, Czech Republic
Dr Jeremy O'Brien, Centre for Quantum Photonics, University of Bristol, England
Prof. Jeremy J. Ramsden, University of Buckingham, England
Prof. Josef Steidl, Czech Technical University, Prague, Czech Republic
Dr E. Clayton Teague, National Nanotechnology Coordination Office (NNCO), Arlington, Virginia, USA
Dr Hans van den Vlekkert, LioniX BV, Enschede, The Netherlands

Nanotechnology Perceptions is published three times a year, in March, July and November
Nanotechnology Perceptions is abstracted/indexed by Inspec, *Chemical Abstracts,* Scopus et al.

nano-ntp.com

Development of Conducting Polymer Metal Oxide Nanocomposite for Ammonia Gas Sensor

Pukhrambam Dipak[1], Dinesh Chandra Tiwari[2]

[1] Department Physics, ITM University Gwalior, India, pukchaokhuman@gmail.com
[2] School of Studies in Physics, Jiwaji University Gwalior, India

Polymer metal oxide nanocomposite (PMON) material is synthesized by the chemical oxidative polymerization. The synthesized nanocomposite is characterized by the Field Emission Scanning Electron Microscope (FESEM), X-ray Diffraction (XRD) and Fourier Transform Infrared Spectroscopy (FTIR). The FESEM images of the PANI (polyaniline) shows the formation of nanofibers of length 200 nm and diameter 50 nm. The FESEM micrograph composite shows that polyaniline is encapsulating the TiO2-Al3+ nanomaterials. The average size of the encapsulation is found to be 300 nm in diameter. The well defined peaks in the XRD spectra confirm the crystalline structure of TiO2-Al3+. Our sensor is able to detect up to 5 ppm concentration of ammonia gas. The sensor is stable up to 100 days and works at room temperature. The slight shift of base line is seen. The sensor is recovered automatically without flushing nitrogen gas. The sensor has the sensitivity of 1.0913 ppm-1.

Keywords: ammonia, metal oxide, nanoparticle, nanocomposite, sensor.

1. Introduction

In the present scenario monitoring different gasses present in an environment is important. Poisonous gasses such as ammonia, methyl isocyanides, dimethyl methylphosphonte (DMMP) etc are harmful for public health. Development of an electronics nose (sensors) for detection of trace amount of poisonous gas became necessary to prevent from fatal accident. The working principle of gas sensor is based on change in resistance (chemi-resistance). The metal oxide, metal halide and conducting polymer gives better detection of ammonia gas at room temperature [1-9]. Conducting polymers are synthesized by various techniques such as chemical polymerization [10-11], electro-chemical polymerization [12], interfacial polymerization [13] and plasma polymerization [14]. An ideal gas sensor should have following characteristics: (i) operation at room temperature; (ii) working in ambient environment and no requirement of oxygen or air supply; (iii) no external stimulus such as

Joule heating or UV illumination for response/recovery; (iv) low detection limit; (v) high sensitivity and reproducibility; (vi) fast response and recovery; (vii) low cost and eco-friendly, etc [15].

Among the various sensor materials, conducting polymer have various advantages, such as less expensive, lighter and can be operated at room temperature. The graphene/PANI (polyaniline) polymer nanocomposite film is deposited on the surface of QCM (quartz crystal microbalance). The sensor shows the good respond toward the ammonia [16]. Flexible PANI/α-Fe2O3 sensors is fabricated and used for detecting the ammonia gas at room temperature. The sensor can detect the ammonia from 5 ppm to 100 ppm [17]. Sadek et.al [18] have developed a polyaniline/In2O3 nanofiber sensor for H2, CO and NO2. Polyaniline is used for various gas detection such as methanol [19], hydrogen [20], ammonia [21], carbon dioxide [22] etc. Polyaniline titanium dioxide nanocomposite sensor is able to sense 23 ppm of ammonia gas [23]. Pawar et.al [24] have also synthesized PANI/TiO2 by chemical oxidative polymerization of aniline with TiO2. This sensor can detect ammonia gas up to 20 ppm. Other conducting polymer nanocomposites such as polypyrrole base sensor are also used for detecting the ammonia gas [10, 25, 26], NO2 [27, 28] and DMMP gas [29].

In the present work we are reporting synthesize of PANI/TiO2-Al3+ nanocomposite for ammonia gas sensing. We are selecting the PANI because it is environmentally stable, easy to synthesize by simple methods and cost effective. The sensor is capable of detecting lowest concentration of 5 ppm of ammonia gas. The prepared sample is characterized by the FESEM, XRD and FTIR.

2. Experimental Method

2.1 Materials

Aniline, sulphuric acid (H2SO4), titanium dioxide(TiO2), hydrochloric acid (HCl) , sodium hydroxide (NaOH), ammonium per sulphate (APS) and aluminium chloride (AlCl3) all are purchased from the Himedia.

2.2 Synthesis of TiO2-Al3+ Nanoparticle (np)

TiO2 and AlCl3 are suspended in 10 M of NaOH and heated for 3 hr at 100oC under reflux with constant stirring. The ph of the suspension is maintained at 5 by adding 1 M of HCl with continuous stirring for 5 hr, so that ion exchange can take place completely forming, NaCl. The solution is allowed to stay for 8 hr and suspended nanoparticles TiO2-Al3+ are settled down. After filtering and washing with deionize (DI) water the nanoparticles are dried at 900C for 6 hr. The synthesized nanoparticles are heated at 4000C for 9 hr.

2.3 Synthesis of PANI/TiO2-Al3+

5 g of TiO2-Al3+ (np) is dispersed in 100 ml of DI water and sonicated for 2 hr. The sonication is followed by stirring. 0.1 M of aniline and 0.1 M of H2SO4 are added and stirred for 1hr. 0.1 M of APS solution is added drop wise to the above solution and kept the solution for 8 hr for polymerization to complete. PANI/TiO2-Al3+ polymer nanocomposite

is filtered and washed with deionize water several times. 0.1 M solution of methanol is used to wash so as to dissolve the unreactived molecules and finally washed with DI water. The nanocomposite is dried at 600C for 2 hr. The sensing material is deposited over the flexible mica sheet. The fabricated sensor is placed inside the gas chamber. The fabricated sensor has the dimension of 2.9 cm X 1 cm and its optical image is shown in Fig. 1. The change in resistance of sensor when exposed to the ammonia gas is recorded by the LCR meter (Hioki-35-32 50).

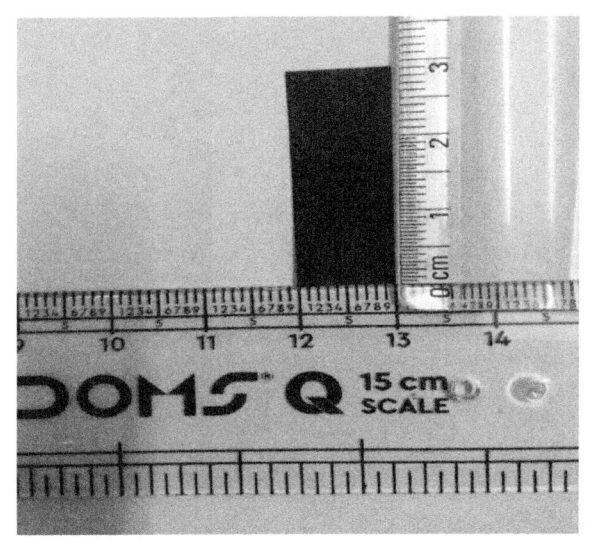

Fig.1 Optical Image of Fabricated Sensor

3. Characterization

The synthesized polymer metal nanocomposite is characterized by the Field Emission Scanning Electron Microscope (FESEM), X-ray Diffraction (XRD) and Fourier Transformation Infrared Spectroscopy (FTIR).

4. Result and Discussion:

4.1 FESEM, XRD and FTIR Studies

The Fig. 2 (a) shows the FESEM image of the PANI. Nanofibers of length of 200 nm and diameter of 50 nm are observed in PANI. Fig. 2 (b) shows the FESEM image of PANI/TiO2-Al3+. In the image the polyaniline is encapsulating the TiO2-Al3+ nanomaterials. The average size of the encapsulation is found to be around 300 nm.

Fig. 2 (c) shows the XRD of PANI and PANI/TiO2-Al3+. The PANI shows single broad peak at 22.750 which infer the amorphous nature of the polymer. In the XRD of polymer nanocomposite the profound peak of PANI is sink and shift to lower angle. This shift in the peak of PANI suggested the formation of the polymer nanocomposite as well as increase is crystallinity of TiO2(np)-Al3+ in polymer nanocomposite is formed [17]. The peaks correspond to various planes such as (1,0,1), (0,0,4), (2,0,0), (1,0,5), (2,1,1) and (2,0,4) are correspond to the TiO2 nanoparticles. The peaks (1,1,1), (2,2,0) and (3,1,1) correspond to Aluminium. The XRD data matches with the JCPDS Card No. 21-1272 for TiO2 and JCPDS 04-0787 for Al respectively.

Fig. 2 (d) shows the FTIR spectra of PANI/TiO2-Al3+ nanocomposite material. The characteristics peaks at 1112.12 cm-1 & 1088 cm-1 are due to C-H plane bending, 1306.15 cm-1 is related to C=N stretching mode, 1486.20 cm-1 & 1483.31 cm-1 correspond to C=C stretching in benzenoid ring, 1576 cm-1 is due to C=C stretching quiniod ring and 794.12 cm-1 correspond to C-H bending vibrations. The peaks found between 600 cm-1 to 450 cm-1 are due to PANI/TiO2-Al3+, which confirms the interaction between PANI and TiO2-Al3+ [30-32, 33].

4.2 Ammonia Gas Sensing:

The ammonia gas sensing set up is shown in the Fig. 3 (a). The sensor performance is studied for different concentration of ammonia. The sensing response (S. R.) of the sensor is calculated by the given formula [10]:

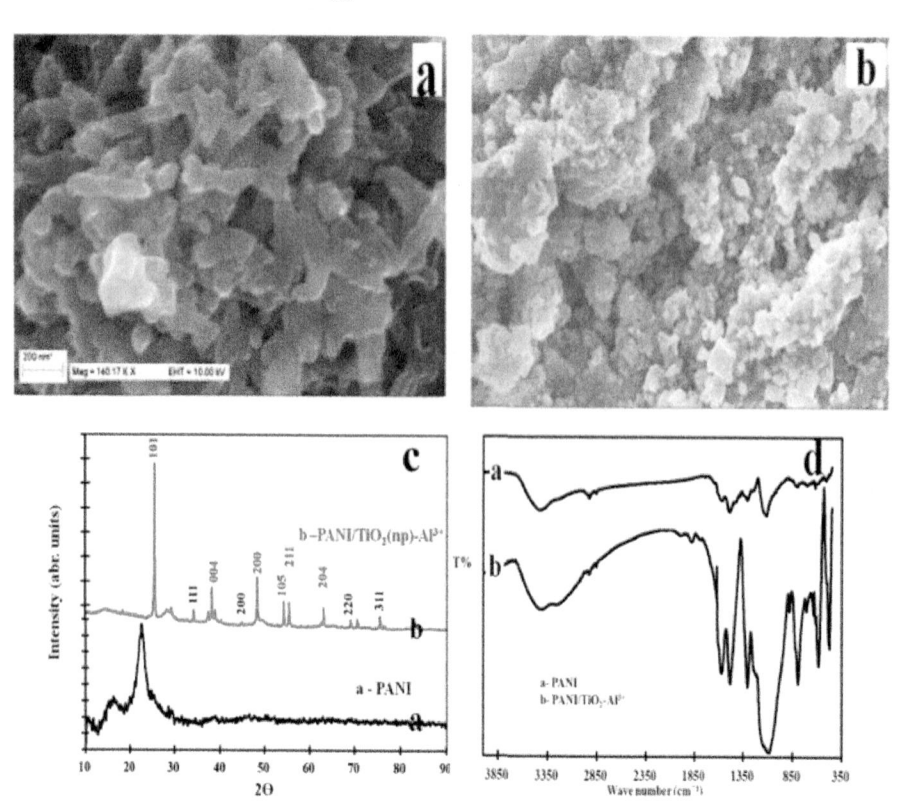

$$S.R.\% = \frac{R_{gas} - R_{air}}{R_{air}} \times 100 = \frac{\Delta R}{R}\% \qquad (1)$$

Fig. 2 (a) FESEM microgram of PANI, (b) FESEM micrograph of PANI/TiO2-Al3+, (c) XRD plot of PANI and PANI/TiO2-Al3+, (d) FTIR plot of PANI and PANI/TiO2-Al3+

Fig. 3 (b) shows the sensor response to ammonia. The sensor is exposed to 5 ppm concentration of ammonia gas and studied the sensing action. The stability of the sensor is checked for various duration of operation up to 100 days. The sensing response of the sensor is fairly constant and the sensor is recovered to the base line without heating. The sensitivity of the sensor remains unchanged from the day one to last day (100 days), the sensing response is found to be 4.25%. But we observed a slight change in the base line of the sensor.

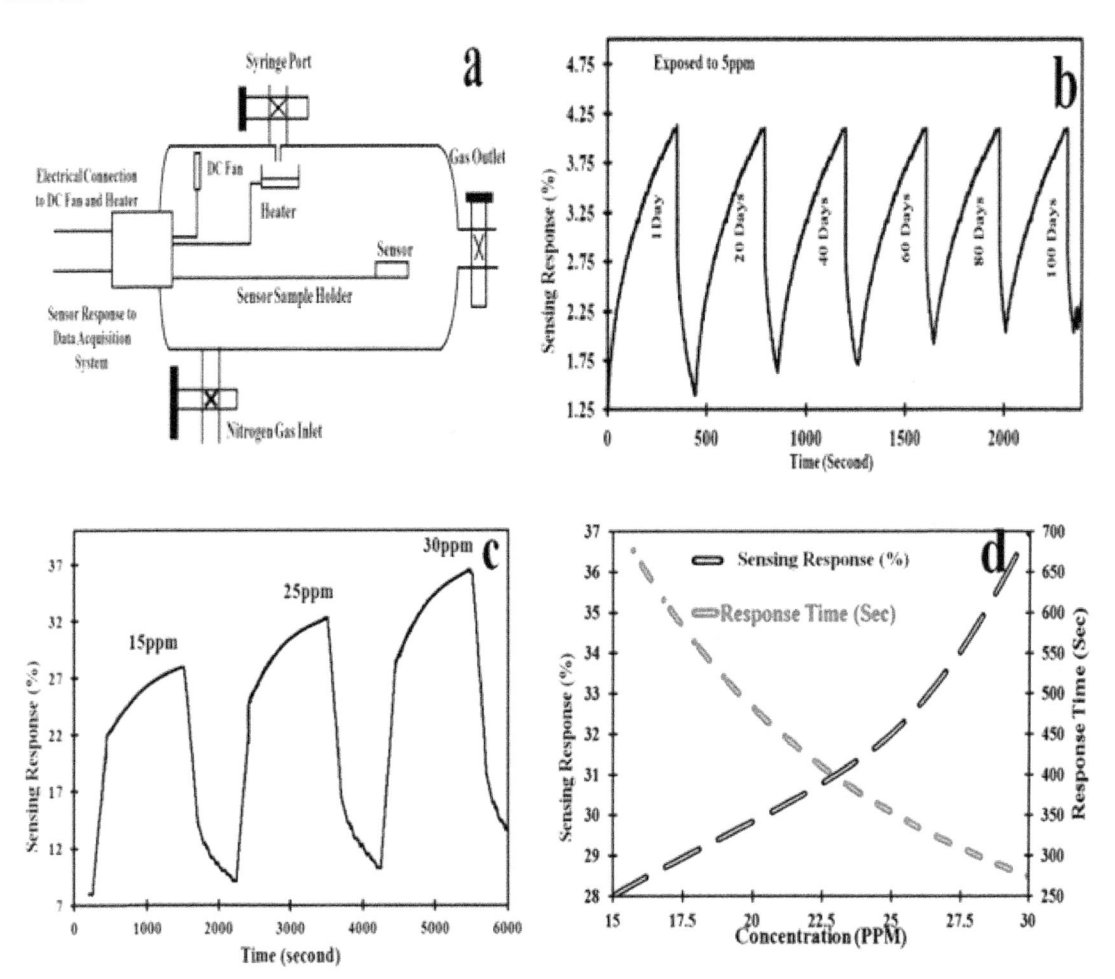

Fig. 3 (a) Gas chamber, (b), (c) sensitivity with time and (d) sensitivity, concentration and response time

Table I comparison with the reported data

Composition	Gas	Lowest Concentration (ppm)	Sensitivity (ppm^{-1})	Authors
PANI/ TiO$_2$-Al^{3+}	NH$_3$	5	1.0913	Present work
PANI-DBSA	NH$_3$	30	-	Yadav et al [34]

PANI/ZnO	NH$_3$	20	~1.2	Das et al [35]
TiO$_2$-SiO$_2$/PANI	NH$_3$	10	~0.12	Pang et al [36]
Cellulose/TiO2/PANI	NH$_3$	10	-	Pang et. al [37]
PANI/TiO$_2$	NH$_3$	23	~0.039	Tai et al [38]
PANI/TiO$_2$	NH$_3$	20	-	Pawar et al [24]
PANI/ZnO	NH$_3$	25	-	Talwar et al [39]

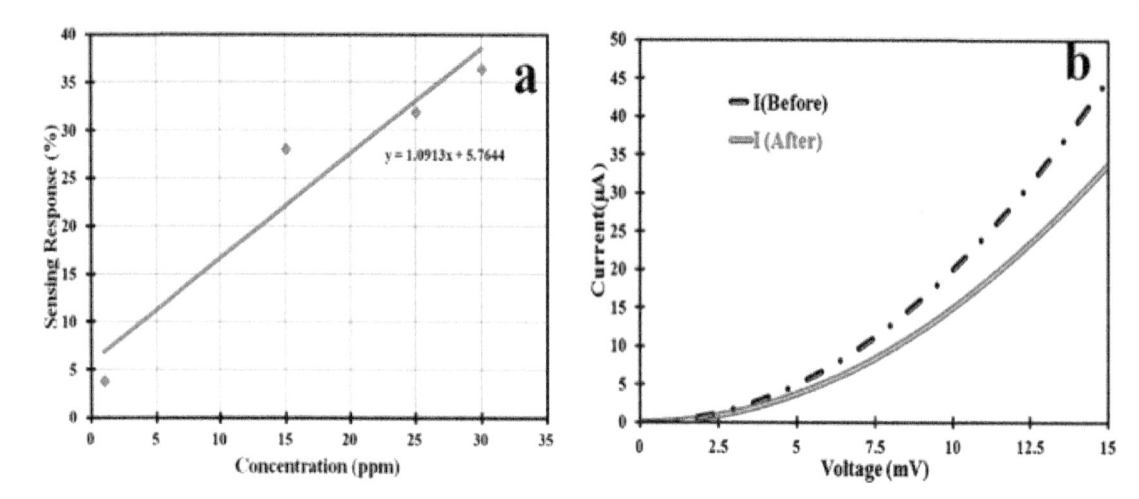

Fig. 4 I-V characteristic of the sensor before and after exposing to the gas

This changed in the base line may be due to the increase in the resistance of the sample caused by the proton donation from the ammonia molecule to the sensor (polymer nanocomposite). The sensor is exposed to different concentrations of gas from 15 ppm to 30 ppm as shown in Fig. 3 (c). After every measurement chamber is flashed with the nitrogen gas. As the concentration of the gas increases, the sensing response of sensor increases from 4.25% to 37% respectively. But the response time of the sensor decreases as the concentration of the ammonia gas increases. Fig. 3 (d) shows the graph between the sensing response and response time with the concentration.

Fig. 4 (a) shows the graph between the sensing responses with concentration, which give the sensitivity of the fabricated gas sensor. The sensitivity of a given sensor is the slope of the graph that plot between the sensing response and concentration (ΔR/ΔC). Thus the sensitivity of the fabricated sensor is found to be 1.0913 ppm-1. Fig. 4 (b) shows the change in resistance of the sensor before and after exposing to ammonia. As PANI in p-type semiconducting material when ammonia molecule came in contact with PANI, the doublet of nitrogen in ammonia looses an electron to the nitrogen of polymer. This electron transfer from ammonia to PANI/TiO2-Al3+ decreases the positive holes density as a result of which the conductance of the sensor decreases and resistance of the sensor increases. Table I shows the comparison between the previous reported work with the present work.

5. Conclusion:

The polymer nanocomposite is synthesized by the chemical polymerization of aniline with TiO2-Al3+ nanomaterials. The FESEM image shows the formation of nanoparticle is encapsulated by polymer. The XRD peaks show the formation of the crystalline structure of the nanomaterials. The sensor can detected the ammonia gas up to 5 ppm concentration. The sensing response of the sensor remained unchanged even after the operation for 100 days. The sensitivity of the sensor is 1.0913 ppm-1.

6. Acknowledgment:

The authors thank MPCST, Bhopal (Project No. A/RD/RP-2/2013-14/214) for providing research grant, Center for Nanosciences and Nanotechnology, IIT Kanpur and Central Facility Instrumentation, Jiwaji University for characterization of the sample.

Conflict of interest: The authors declare that we have no conflict of interest. We have already acknowledged the funding agency in Acknowledgment section.

References

1. M. Sahm, A. Oprea, N. Barsan, U. Weimar, Water and ammonia influence on the conduction mechanisms in polyacrylic acid films, Sens. Actuators B 127, 204–209 (2007).
2. B. H. King, A. Gramada, J. R. Link, M. J. Sailor, Internally referenced ammonia sensor based on an electrochemically prepared porous SiO2 photonic crystal, Adv. Mater. 19, 4044–4048 (2007).
3. S. G.Ansari, Z. A.Ansari, H. K, Seo, G. S. Kim, Y. S. Kim, G. Khang, H. S. Shin, Urea sensor based on tin oxide thin films prepared by modified plasma enhanced CVD, Sens. Actuators B 132, 265–271 (2008).
4. L. R. Narasimhan, A. Goodman, N. Patel, Correlation of breath ammonia with blood urea nitrogen and creatinine during hemodialysis, Proc. Natl. Acad. Sci. 98, 4617–4621 (2001).
5. M. B. Pushkarsky, M. E. Webber, O. Baghdassarian, L. R. Narasimhan, C. K. N. Patel, Laser-based photoacoustic ammonia sensors for industrial applications, Appl. Phys. B 75, 391–396 (2002).
6. B. Timmer, W. Olthuis, V. D. Berg, Ammonia sensors and their applications—a review, Sens. Actuators B 107, 666–677 (2005).
7. C. Y. Shen, S. Y. Liou, Surface acoustic wave gas monitor for ppm ammonia detection, Sens. Actuators B 131, 673–679 (2008).
8. A, Maity, A. K. Raychaudhuri, B. Ghosh, High sensitivity NH3 gas sensor with electrical readout made on paper with perovskite halide as sensor material, Scientific Reports (2019) 9:7777 | https://doi.org/10.1038/s41598-019-43961-6.
9. S. Christie, E. Scorsone, K. Persaud, F. Kvasnik, Remote detection of gaseous ammonia using the near infrared transmission properties of polyaniline, Sens. Actuators B 90, 163–169 (2003).
10. D. C. Tiwari, Priyanka Atri, Rishi Sharma, Synthesis of reduced graphene oxide nanoscrolls embedded in polypyrrole matrix for supercapacitor applications, Synth. Metal, 21-28 (2017).
11. D. C. Tiwari, P. Dipak, S. K. Dwivedi, T. C. Shami, P. Dwived, PPy/TiO2(np)/CNT polymer nanocomposite material for microwave absorption, J Mater Sci: Mater Electro, 29, 1643-1650 (2018).

12. B. Wang, J. Tang, F. Wang. Synth. Metal, Vol 18, Issues 1–3, 323-328 (1987).
13. C. J. Michaelson, A. J. Mc Evoy, Interfacial polymerization of aniline, J. Chem. Soc. Commun., 1994.
14. G. J. Cruz, J. Morales, R. Olavo, Films obtained by plasma polymerization of pyrrole, Thin Solid Film Vol 342, Issues 1–2, 26, 119-126 (1999).
15. S. Pandey, Gopal K. Goswami, K. K. Nanda, Nanocomposite based flexible ultrasensitive resistive gas sensor for chemical reactions studies, Sci. Rep. 3, 2082; DOI: 10.1038/srep02082 (2013).
16. Z. Wu, X. Chen, S. Zhu, Z. Zhou, Y. Yao, W Quan, B. Liu, Enhanced sensitivity of ammonia sensor using graphene/polyaniline nanocomposite, Sens. and Actuators B 178 (2013), 485– 493.
17. D. K. Bandgar, S. T. Navale, M. Naushad, R.S. Mane, F. J. Stadler, V. B. Patil, Ultra-sensitive polyaniline-iron oxide nanocomposite room temperature flexible ammonia sensor, RSC Advances, 5 (2015) 68964-68971.
18. A. Z. Sadek, W. Wlodarski, K. Shin, R. Bkaner, K. Kalantar-zadeh, A layered surface acoustic wave gas sensor based on a polyaniline/In2O3 nanofibre composite, Nanotechnology17 (2006) 4488–4492.
19. M. G. H. Meijerink, D.J. Strike, N.F. Rooij, M. Koudelka-Hep, Reproducible fabrication of an array of gas-sensitive chemo-resistors with commercially available polyaniline, Sens. Actuators B Chem. 68 (2000) 331-334.
20. C. K. Tan, D. J. Blackwood, Interactions between polyaniline and methanol vapour, Sens. Actuator B 71 (2000) 184.
21. C. Conn, S. Stephen, A. T. Baker, J. Unsworth, A Polyaniline-Based Selective Hydrogen Sensor, Electroanalysis 10 (1998) 1137-1141.
22. P. Kiattibutr, L. Tarachiwin, L. Ruangchuay, A. Sirivat, J. Schwank, Electrical conductivity responses of polyaniline films to SO2-N2 mixtures: Effect of dopant type and doping level, React. Funct. Polym. 53 (2002) 29.
23. H. Tai, Y. Juang, G. Xie, J. Yu, and X. Chen, Fabrication and gas sensitivity of polyaniline–titanium dioxide nanocomposite thin film, Sens. Actuators B, 125 (2007) 664–650.
24. S. G. Pawar, M. A. Chougule, S. L. Patil, B. T. Raut, P. R. Godse, Shashwati Sen, V. B. Patil, Room Temperature Ammonia Gas Sensor Based on Polyaniline-TiO2 Nanocomposite, IEEE Senors J., Vol. 11, No. 12, (2011).
25. Q. Ameer, S. B. Adeloju, Polypyrrole-based electronic noses for environmental and industrial analysis, Sens. Actuators B, 106(2005)541-552.
26. C. Mahajan, P. Chaudari, S. Mishra, Study of structural, optical and magnetic properties of Ni-doped ZnO nanoparticles synthesized by co-precipitation method, J. of Mater. Science: Materials in Electronics 29 (2018), 10, 8039–804.
27. S. Radhakrishnan, S. D. Deshpande, Effect of copper substitution on the magnetic and electrical properties of nanocrystalline nickel-zinc ferrites, Mat. Sci. Lett, 48(2001)144.
28. R. A. Naikoo, S. U. Bhat, M. A. Mir, R. Tomar, W. A. Khanday, P. Dipak, D. C. Tiwari, Polypyrrole and its composites with various cation exchanged forms of zeolite X and their role in sensitive detection of carbon monoxide, RSC Adv., 2016, 6, 99202.
29. D. C. Tiwari, R. Sharma, K. D. Vyas, M. Boopathi, V. V. Singh, P. Pandey, Electrochemical Incorporation of Copper Phthalocyanine in Conducting Polypyrrole for the Sensing of DMMP, Sens. and Actuators B 151 (2010) 256–264.
30. P. Sharma, J. Yeo, D. K. Kim, C. H. Cho, Organic additive free synthesis of mesoporous

naoncrystalline NaA zeolite using high concentration inorganic precursors, J. Mater. Chem., 22 (2012) 2838-2843.

31. X. Feng, Y. Zhang, Z. Yang, N. Chen, Y. Ma, X. Liy, X. Yang, W. Hou, Self-degradable template synthesis of polyaniline nanotubes and their high performance in the detection of dopamine, J. Mater. Chem., 1 (2013) 9775-9780.

32. F. Xu, G. Zheng, D. Wu, Y. Liang, Z. Li, R. Fu, Improving electrochemical performance of polyaniline by introducing carbon aero gel as filler, Phys. Chem. Chem. Phys., 12 (2010), 3270-3275.

33. P. Dipak, D. C. Tiwari, A. Samadhiya, N. Kumar, Th. Biswajit, P. A. Singh, R. K. Tiwari, Synthesis of polyaniline (printable nanoink) gas sensor for the detection of ammonia gas, J. Mater. Sci: Mater. Electron, 31 (2020) 22512-22521.

34. A. Yadav, A. Agarwal, P. B. Agarwal, P. Saini, Ammonia Sensing by PANI-DBSA Based Gas Sensor Exploiting Kelvin Probe TechniqueJ. of Nanoparticles, 2015, Article ID 842536, 6 pages,doi.org/10.1155/2015/842536.

35. M. Das, D. Sarkar, One-pot synthesis of zinc oxide - polyaniline nanocomposite for fabrication of efficient room temperature ammonia gas sensor, Ceramics International, 43(14), 11123–11131.doi:10.1016/j.ceramint.2017.05.159.

36. Z. Pang, J. Yu, D. Li, Q. Nie, J. Zhang, Q. Wei, Free-standing TiO2–SiO2/PANI composite nanofibers for ammonia sensors, J. of Mater. Sc.: Materials in Electronics (2017) https://doi.org/10.1007/s10854-017-8287-2.

37. Z. Pang, J. Fu, L. Luo, F. Huang, Q. Fabrication of PA6/TiO2/PANI composite nanofibers by electrospinning–electrospraying for ammonia sensor, Colloids and Surfaces A: Physicochemical and Eng. Aspects 461 (2014) 113–118.

38. H. Tai, Y. Jiang, G. Xie, J. Yu, X. Chen, Fabrication and gas sensitivity of polyaniline–titanium dioxide nanocomposite thin film, Sens. and Actuators B 125 (2007) 644–650.

39. V. Talwar, O. Singh, R. C. Singh, ZnO assisted polyaniline nanofibers and its application as ammonia gas sensor, Sens. and Actuators B 191 (2014) 276– 282.

Performance Evaluation of an Efficient ALU

Mary Sajin Sanju.I[1], M. Vadivel[2]

[1] *Research Scholar, Sathyabama Institute of Science and Technology, Chennai, India*
[2] *Vidya Jyothi Institute of Technology, Hyderabad, India*

The arithmetic logic unit (ALU) is an important building block in many applications such as microprocessors, digital signal processors (DSPs) and image processing. Power efficiency may be a general concern in VLSI design. This paper presents delay-time optimization of a 4-bit ALU designed using the full-swing gate diffusion input (GDI) technique. An efficient ALU was designed and simulated by using HSPICE tool using 130nm technology. The modified ALU gives better performance in terms of power, delay and energy and can be used for high-speed and low-power applications.

Keywords: Arithmetic Logic Unit (ALU); Gate Diffusion Input (GDI), Full-Swing GDI (FS-GDI).

1. Introduction

Power utilization and area occupied are the most pressing issues within the semiconductor industry, which have motivated great research effort to attenuate power consumption and concomitantly the area of VLSI circuits. Severely limited power is available for the portable electronic devices heavily used day-to-day worldwide. These devices are simultaneously low-power and high-speed.

Compared to CMOS and pass transistor logic (PTL) techniques, the gate diffusion input (GDI) technique allows improvement in power consumption, propagation dissipation delay and area occupied by VLSI digital circuits. [1]

In 2019 Mahmoud Aymen Ahmed designed an ALU using the full-swing GDI technique[2]. Simulations carried out in Cadence Virtuoso using 65 nm TSMC processes with a supply voltage of 1.2 volts and a frequency of 125 MHz revealed improvement in delay time and overall energy of the optimized ALU design. Using the GDI technique allowed improvements in power, propagation delay and VLSI circuit area compared to the CMOS and pass transistor logic (PTL) techniques. But it was proposed for fabrication in twin-well CMOS or silicon-on-insulator (SOI) processes and, similarly to PTL, the GDI gates suffered from reduced voltage swings at their outputs due to threshold drops [3]. This increased static power dissipation and caused performance degradation.

Alex F. Kirichenko designed and tested a parallel 8-bit ERSFQ ALU, comprising 6840 Josephson junctions, in 2019 [4] . The ALU design employed wave-pipelined instruction execution and featured modular bit-slice architecture that is easily extendable to any number of bits and adaptable to current recycling. A carrier signal synchronized with asynchronous instruction propagation provided the wave-pipeline operation of the ALU [5]. Its instruction set consists of 14 arithmetical and logical instructions. It was designed and simulated for operation up to 10 GHz clock rate and embedded in a shift register-based high frequency test bed with an on-chip clock generator to allow for comprehensive high frequency testing for all possible operands. It was fabricated with MIT Lincoln Laboratory's 10 kA/cm2 SFQ5ee fabrication process featuring eight Nb wiring layers and the high kinetic inductance layer needed for ERSFQ technology.

Parth Khatter designed an ALU using basic reversible gates in 2018 [6]. Two designs for both the arithmetic and the logical unit were proposed and one design for the control unit was also put forward. These designs were integrated together to give four complete ALU designs. Each proposed design was implemented in Verilog HDL using Xilinx ISE Suite 14.1 software to verify its functionality[7]. The proposed designs were compared with each other and their existing counterparts based on quantum cost, garbage outputs and ancillary inputs [8]. The simplicity and the reduced quantum cost of the proposed ALU designs make them ideal candidates to be used as modules in quantum computers. The proposed ALUs extend their applications to cryptography, machine learning and nanotechnology [9]. The technological advances have made integration of thousands of millions of transistors on single die possible, which has given the designers the flexibility and freedom of putting more and more functionality on the same die. This has, however, resulted in increased power consumption, which has motivated a plethora of techniques for dealing with it.

K. Pandiammal designed digital circuits at the nanoscale in the same year (2018). [10]An 8-bit QCA based reconfigurable 1-bit ALU was proposed using clock zone-based crossover (CZBCO). The ALU unit was designed to perform four arithmetic and logical operations, viz. binary addition, logical AND, OR and EXOR. The multigates (MGs) used in the EXOR operation [9]are reused for logical AND and OR their number reduces to two in 1-bit ALU design. The proposed ALU reduced energy dissipation by 54.5% and minimized QCA cell utilization by 43.5% when compared to existing devices.

Again in 2018 Alexis Ramos designed ALU with the use of microprocessors in space missions in mind, implying that they should be protected against the effects of cosmic radiation. Commonly this objective has been achieved by applying modular redundancy, which provides good results in terms of reliability but obviously significantly increases the number of used resources. [11] Because of that, new protection techniques have appeared, trying to establish a trade-off between reliability and resource utilization.

An application-based methodology is proposed by A .Ramos in 2019 to protect a soft processor implemented in an SRAM-based FPGA against the effect of soft errors. This is done creating a library of adaptive protection configurations, based on the profiling of the application. This hardware configuration library, combined with the reprogramming capabilities of the FPGA, helps to create an adaptive protection for each application. Two partial triple modular redundancy (TMR) configurations for the ALU are presented as an example of this methodology, in which TMR preserves the error present in the functional unit [13]. Those components leading to failure are identified and increased reliability is achieved. The proposed scheme has been tested in a RISC-V soft processor. A fault injection campaign has been carried out to test its reliability.

There is a need to reduce overall design cost, a digital system is implemented by considering the constraints of power, area, speed and cost of the internal logic blockk [14] .The proposal aimed to implement different functionalities over the same resources with a time multiplexing concept. Depending on the requirement a suitable adder or a multiplier could be switched in without using extra resources [15] . An effort was made to explore the possibility of run-time reconfiguration as applied to 4-bit ALU design, hence leading to an efficient utilization of the resources available on the device. Design comparison indicating resource utilization and power consumption with and without partial reconfiguration was given. The design was executed using XILINX Vivado, which was also used by Jumuna in 2019 for an ALU implemented on FPGAs to analyse the design parameters, with the main design objective being to develop algorithms to achieve an efficient utilization of the available hardware [16]. The measures of the efficiency of an algorithm are speed improvement, less power consumption and better utilization of the ALU [17].

Simulated and synthesized parameters of ALUs by using VERILOG on Xilinx is proposed by Amrit Kumar Panigrahi (2019), an attempt was made to demonstrate reconfigurable built-in self-test (BIST) logic, which detects faults across the ALU block mapped on the FPGA. Proposed work aims at detecting stuck-at faults occurring at the internal blocks of ALU. Stuck at fault is a particular fault model used in automatic test pattern generation. Test patterns are generated through LFSR and outputs are analyzed using MISR(Multiple input signature register) block. MISR block is is multiple output digital system used for testable design purpose which accelerates the task of compressing multiple input data streams into one signature in the testing task. Through this proposed work an effort is made to verify the design also using Questasim simulator. The design is been executed using XILINX Vivado IDE. Number of faults detected is 80 out of 98 with code coverage of 94.2% and the total power consumed by the proposed design is 1.06 watts. Kp Shashikala (2019) designed an Arithmetic Logic Unit (ALU) by taking advantage of the concept of dual mode logic (DML) technique has been implemented. ALU is one of the most significant component of any computing system be it microprocessors, embedded structures or any other computational device. In this, ALU consists of 4x1 multiplexer, 2- input and unit, 2-input or unit, 2-input exor unit and a full adder designed to implement logic operations, such as and, or, exor and arithmetic operation of addition using a full adder. DML technique has been used for designing of multiplexer, full adder, and unit, or unit and exor unit which are then associated to realize the DML based ALU. Power of static CMOS ALU and delay of domino ALU architecture were calculated which were then compared with static mode and dynamic mode of DML based ALU respectively. This is designed and simulated using Mentor Graphics Pyxis Schematic Tool with 1.8 V supply voltage and 180 nanometre (nm) technology.

2. Proposed system

The proposed Full Adder design shows a detailed circuit of the proposed Full adder. Sum output (O/p) is constructed by two cascaded XNOR modules/gates. Signal B is applied to the weak inverter comprised of Q3, Q4. Input (I/p) and Output (O/p) signals (B and B' respectively), of this weak inverter are used to construct the controlled inverter with Q3 and Q4 Transistors. To overcome this swing degradation problem, pass transistor Q5 and Q6 are used. P-MOS pass transistor (Q5) and N-MOS pass transistor (Q6) for Strong. The output (O/p) of the first XNOR module is applied as input (I/p) to the second XNOR module for complete SUM function.

The proposed 4-Bit ALU designed using 65nm TSMC CMOS process, The simulations were done using the SPECTRE based Cadence Virtuoso simulator with a power supply 1.2V, and a clock frequency of 125 MHz, the size of PMOS is twice that of the NMOS transistor (W/L) =240 nm/60 nm and (W/L) =120 nm/60 nm for best power and delay performance. Using A=1100, B=0101 as test inputs. The proposed design compared to the previous design in [5] in terms of power consumption, delay, energy and transistor count. Simulation results for the proposed 4-bit ALU are shown below.

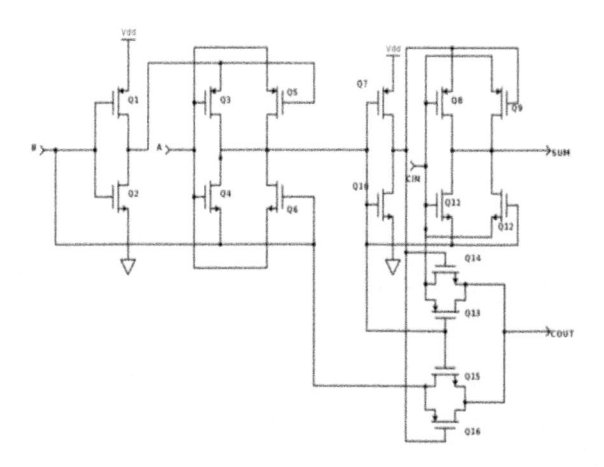

Fig:1 Proposed System

In This work delay time of a 4-Bit ALU designed using the full-swing GDI technique optimized and reduced by 22.3% compared to the previous design, while maintaining full-swing operation. Hence the energy of the 4-bit ALU reduced by 21.2%. The proposed design consists of 294 transistors and operates under 1.2V supply voltage and Frequency of 125 MHz, based on the results, it can be concluded that the proposed 4-bit ALU in full-swing GDI technique is suitable for low energy high-speed VLSI applications. Further study in this work would be using the 4-bit ALU as a building block to implement 8-bit and 16-bit ALU. The proposed 4-Bit ALU designed using 65nm TSMC CMOS process, The simulations were done using the SPECTRE based Cadence Virtuoso simulator with a power supply 1.2V, and a clock frequency of 125 MHz, the size of PMOS is twice that of the NMOS transistor (W/L) = 240 nm/60 nm and (W/L) =120 nm/60 nm for best power and delay performance. Using A=1100, B=0101 as test inputs. The proposed design compared to the previous design in [5] in terms of power consumption, delay, energy and transistor count. Simulation results for the proposed 4-bit ALU are shown. An adder is a digital circuit that performs addition of numbers. The half adder adds two binary digits called as augend and addend and produces two outputs as sum and carry; XOR is applied to both inputs to produce sum and AND gate is applied to both inputs to produce carry. The full adder adds 3 one bit numbers, where two can be referred to as operands andone can be referred to as bit carried in. And produces 2-bit output, and these can be referred to as output carry and sum. Then we can see that Multiplexers are switching circuits that just switch or route signals through themselves, and being a combinational circuit they are memory less as there is no signal feedback path. The multiplexer is a very useful electronic circuit that has uses in many different applications such as signal routing, data communications and data bus control applications. The advantage is that only one serial data line is required instead of multiple parallel data lines. Therefore, multiplexers are sometimes referred to as "data selectors", as they select the data to the line.

3. Simulation result

The existing and proposed architectures are simulated using MODELSIM. Arithmetic logic unit is an essential building block in many applications such as microprocessors, DSP and image processing while power efficiency is a general concern in VLSI Design. This presents delay time optimization of 4 bit ALU designed. Simulation results revealed improvement in Delay time and overall energy of the optimized ALU design. Output of the existing ALU and proposed ALU are analysed from the results and it is concluded that the proposed architecture reduces the delay compared with the existing system. The Synthesis is done by using HSPUI A. 5.1 SIMULATION WAVE FORM FOR EXISTING SYSTEM

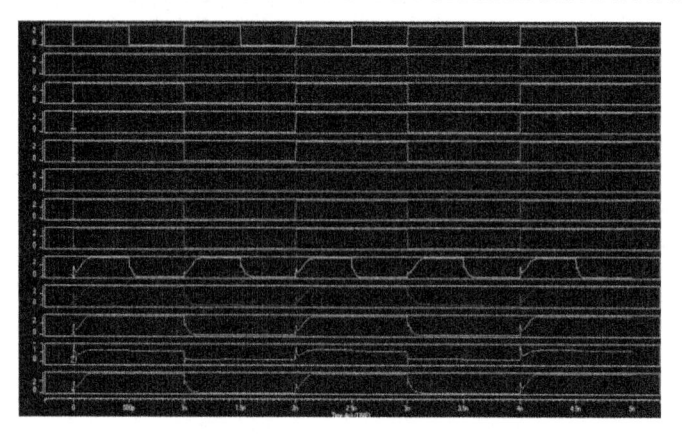

Fig:2 Simulation wave form for Existing ALU

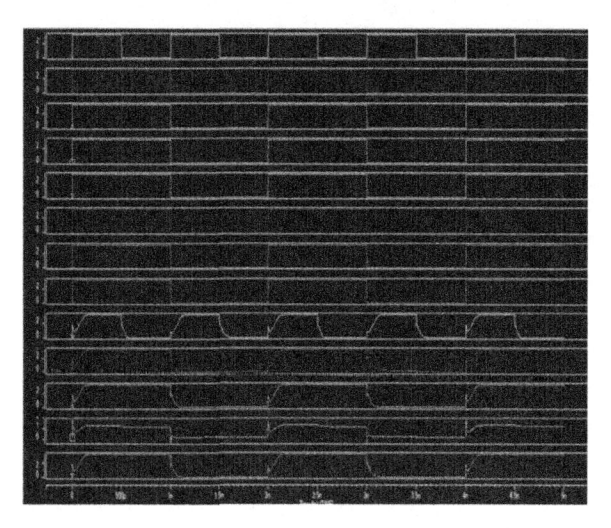

Fig:3 Simulation Wave Form for Proposed ALU

The output waveform of the existing ALU shown in fig:2, each function presented in one nanosecond of time with the same order. The output of each function is clearly described by the respective signals shown in the above figure. The proposed ALU simulated wave is shown in Fgure 3. The output waveform of the proposed ALU and each function is presented in one nanosecond of time with the same order. The output of each function clearly shows improvement compared to the existing ALU.

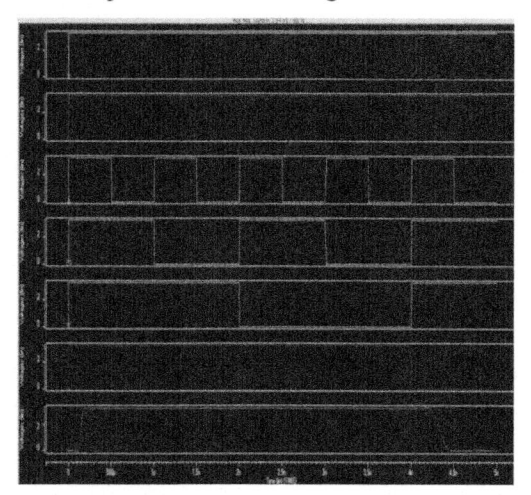

Fig. 4. Simulation waveform for 4x1

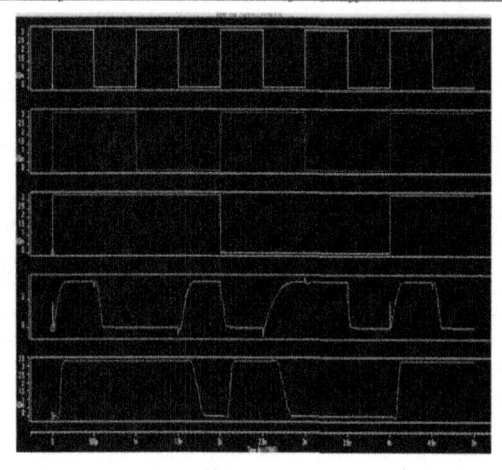

Fig. 5. Simulation Waveform for an adder

The simulated waveform of the Adder shown in Fig: (4 & 5), each function presented in one nanosecond of time with the same order. The output of the adder circuit is described with respective to the above signals.

Table 1. Comparative analysis of the existing and proposed systems.

Design	Average power/nW	Peak power/nW	No of transistors
Existing system	0.4091	0.5641	12
Proposed system	0.1541	4.398	20

DESIGN	AVG POWER	PEAK POWER
4x1 multiplexer	1.193e-05	4.732e-04
OR gate	1.715e-05	7.639e-03
Adder Circuit	1.059e-04	1.222e-02
XOR gate	3.476e-05	3.469e-03
AND gate	1.660e-05	3.447e-03

A comparative analysis is made on the existing and the proposed one which is shown in the above Table:1.The analysis and simulation states that the proposed ALU design decreased Delay time by 22.3% (15.6ps) and improved Energy by 21.2% at a slight increase in Transistor count (8 Transistors and in Power consumption (390nW). The average power decreased by 2.55e-04 when compared to the previous design whereas the peak power has been decreased by 1.243e-03.

4. Conclusion

The delay time of a 4-bit ALU designed using the full-swing GDI technique optimized and reduced by 22.3% compared to the previous design as shown above, while maintaining full swing operation. Hence the energy of the 4-bit ALU reduced by 21.2%. The proposed design consists of 294 transistors and operates under 1.2V supply voltage and frequency of 125MHz, based on the results carried out, it can be concluded that the proposed 4-bit ALU in full swing GDI technique is suitable for low power high-speed Very LargeScale Integrated Applications. Further study during this work would be using the 4-bit ALU as a building block to increment 8 bit and 16-bit ALU.An arithmetic logic unit (ALU) is a digital electronic circuit that performs arithmetic and bit wise logical operations on integer binary numbers. Arithmetic and logical unit is a fundamental building block of many types of computing circuits, including the central processing unit (CPU) of computers, FPUs, and graphics processing units. Arithmetic Logic Unit (ALU) is an essential building block in many applications such as microprocessors, DSP, and image processing, while power efficiency is a general concern in VLSI design. The design and simulation of an efficient ALU are accomplished in HSPICE tool based on 130nm technology. From the findings modified Arithmetic and Logical Unit gives better performance in terms of power, Delay and energy which can we used for high speed and low power application

References

[1] V. Yuzhaninov, I. Levi and A. Fish, Design flow and characterization methodology for dual mode logic. IEEE Access 3 (2016) 3089–3101.

[2] Mahmoud Aymen Ahmed1 , M. A. Mohamed El-Bendary2 "Delay Optimization of 4-Bit ALU Designed in FS-GDI Technique" 2019 International Conference on Innovative Trends in Computer Engineering (ITCE'2019), Aswan, Egypt, 2-4 February 2019.

[3] A. Morgenshtein, A. Fish, and I. A. Wagner, Gate-diffusion input (GDI): A power-efficient method for digital combinatorial circuits. IEEE Trans Very Large Scale Integration (VLSI) Systems 10 (2002) 566–581.

[4] A.F.Kirichenko.I.V.Vernik.M.Y.Kamkar,J.Walter,M.Miller" ERSFQ 8-bit Parallel Arithmetic Logic Unit"2019, IEEE trans.29(5),13022407,Aug(2019)DOI:10.1109/TASC2019.2904484.

[5] A. Kaizerman, S. Fisher and A. Fish, Sub threshold dual mode logic. IEEE Trans Very Large Scale Integration (VLSI) Systems 21 (2018) 979–983.

[6] P. Khatter, N. Pandey and K. Gupta, "An Arithmetic and Logical Unit using Reversible Gates10.1109/TASC2019.2904484.," 2018 International Conference on Computing, Power and Communication Technologies (GUCON), 2018, pp. 476-480, doi: 10.1109/GUCON.2018.8675034.

[7] Shiksha and K.K. Kashyap, High speed domino logic circuit for improved performance. 2014 Students Conference on Engineering and Systems (SCES), 28–30 May, pp. 1–5.

[8] K. Vinay Kumar, F. Noorbasha, B. Shiva Kumar and N.V. Siva Rama Krishna, Design of an efficient ALU using low-power dual-mode logic. Intl J. Engng Res. Applications (IJERA) 4 (May 2014) 81–84.

[9] M.M. Mano and C.R. Kime, Logic and Computer Design Fundamentals. Harlow: Pearson Education (2015).

[10] K. Pandiammal and D. Meganathan, "Design of 8 bit Reconfigurable ALU Using Quantum Dot Cellular Automata," 2018 IEEE 13th Nanotechnology Materials and Devices Conference (NMDC), 2018, pp. 1-4, doi: 10.1109/NMDC.2018.8605892.

[11] G. Tang, K. Takata, M. Tanaka, A. Fujimaki, K. Takagi and N. Takagi, 4-bit bit slice arithmetic logic unit for 32-bit RSFQ microprocessors. IEEE Trans. Appl. Superconductivity 26 (2016) 1300106.

[12] A. Ramos, R. G. Toral, P. Reviriego and J. A. Maestro, "An ALU Protection Methodology for Soft Processors on SRAM-Based FPGAs," in IEEE Transactions on Computers, vol. 68, no. 9, pp. 1404-1410,1 Sept. 2019, doi: 10.1109/TC.2019.2907238.

[13] Y. Ando, R. Sato, M. Tanaka, K. Takagi and N. Takagi, 80-GHz operation of an 8-bit RSFQ arithmetic logic unit. Proc. 15th Intl Superconductive Electronics Conf. (ISEC), Nagoya (2015), pp. 1–3.

[14] M. A. Ahmed and M. A. Abdelghany, Low-power 4-bit arithmetic logic unit using full-swing GDI technique. Proc. Intl Conf. on Innovative Trends in Computer Engineering (ITCE 2018), pp. 193–196.

[15] V. Dubey and R. Sairam, An arithmetic and logic unit optimized for area and power. 4th Intl Conf. Advanced Computing & Communication Technologies (ACCT'14), pp. 330–334 (2014)

[16] T. Filippov, M. Dorojevets, A. Sahu, A. Kirichenko, C. Ayala and O. Mukhanov, 8-bit asynchronous wave-pipelined RSFQ arithmetic logic unit. IEEE Trans. Appl. Superconductivity 21 (2011)

[17] P. Khatter, N. Pandey and K. Gupta, "An Arithmetic and Logical Unit using Reversible Gates10.1109/TASC2019.2904484.," 2018 International Conference on Computing, Power and Communication Technologies (GUCON), 2018, pp. 476-480, doi: 10.1109/GUCON.2018.8675034.

Optimizing the Parameters of diodes based on Gallium Sulfide Monolayer and Improving its Performance by Injecting Impurities and Introducing Defects

Mojtaba Moshtael[1], Maryam Nayeri[1*], Fatemeh Ostovari[2]

[1] Department of Electrical Engineering, Yazd Branch, Islamic Azad University, Yazd, Iran.
[2] Department of Physics, Yazd University, Yazd, Iran.
* Corresponding Email: nayeri@iauyazd.ac.ir

The gallium sulfide (GaS) monolayer is a potential two-dimensional semiconductor material with exceptional characteristics. GaS has a light gap of approximately 2.48 eV. However, it is indirect meaning the conduction band edge and finesse does not reside at a point in the wave vector's space. In this research, density functional theory-based calculations were used to study the diode properties of GaS nanowires. First, the parameters relevant to the analysis, such as the limit energy and the inverting spatial network, are optimized ,and then the GaS unit cell is optimized based on these parameters. The ideal grid vector and link length were used to complete the calculations. The electronic properties of the GaS transport channel with various impurities were investigated using state density calculations as well. The electronic transport of the GaS transport channel with impurities from different atoms was then estimated and examined using a current-voltage diagram. The diode characteristics of GaS nanowires were investigated using calculations based on modified density functional theory in this contribution. The GaS unit cell is optimized for this purpose by first optimizing the parameters connected with the computations, such as limit energy and inverted space meshing, and then employing these parameters. The best grid vector and connection length were utilized to continue the computations. Subsequently, the density of states calculations were used to investigate the electronic properties of the GaS transport channel with various impurities. Later, the electronic transport of the GaS transport channel with impurities from different atoms was calculated and investigated by calculating the current-voltage diagram.

Keywords: Gallium sulfide, Diode, Monolayer, Defects.

1. Introduction

2D materials hold great promise for developing new nanodevices for future applications because of their fascinating electronic, mechanical, optical, and thermal properties, which can be attributed to their small dimensionality and quantum confinement effect. These nanostructures have novel properties that are superior to their solid counterparts [1]. 2D nanosheets with hexagonal structure such as graphene [2], a hexagonal boron nitride (h-BN) [3], silicene [4, 5], stanine [6], phosphors [7, 8] ,and transition metal dichalcogenides (TMDs) [9-12] have attracted extensive research efforts in recent years due to their excellent performance in nanoelectronic devices. The novel properties of these 2D materials hold promise for numerous industrial applications such as optoelectronics, photodetectors, catalysts, superconductors, and spintronic films [13]. However, the usefulness of these materials is limited by some shortcomings, such as the lack of a bandgap in graphene and the relatively low mobility in some TMDs [14]. This has motivated us to continue the search for other 2D materials that can demonstrate their properties and lead to an improvement in specific performance. Recently, another class of 2D materials, metal chalcogenides, was discovered and received much attention. These layer materials generally have the chemical formula MX, where M and X belong to group IIIA and IVA, respectively. GaS nanosheet has been successfully isolated by micromechanical cleavage technology [15, 16]. Monolayer gallium sulfide has been used in photodetectors, photoelectric devices, electrical sensors, devices that emit near blue light [16-19], and field-effect transistors (FETs). The mobility of 0.1 cm2 / V.S. was determined by David et al. [20] measured in ultra-thin GaS bottom-gate transistors. In addition, the piezoelectric coefficients of GaS monolayers are in the same order of magnitude as those of discovered two-dimensional (2D) piezoelectric materials such as boron nitride (BN) and MoS 2 monolayers [21].Gallium sulfide is an indirect bandgap semiconductor that interferes with optoelectronic devices that emit light and use photodetectors, and lasers. Since stresses are crucial in the development of band gaps in 2D structures [22-31], it would be exciting to investigate the influence of stresses on the band structure of GaS. If the energy difference between the indirect and direct band gaps is slight, a transition from indirect to direct bandgap is conceivable. Monochalcogenes are two-dimensional structures made of the thirteenth and sixteenth elements of the periodic table. The stoichiometry of these structures is MX, where M is an element from the thirteenth group and X is from the sixteenth group of the periodic table. The term chalcogen belongs to the elements of the sixteenth group, and since the unit cells of the structures in question are composed of the elements of this group and the elements of the thirteenth group in a ratio of 1 to 1, they are called monochalcogenes or single chalcogens. These compounds include GaS, GaSe, InS ,and InSe. The two-intermediate metal chalcogens, phosphorus, arsenic ,and monochalcogenes ,all have a bandgap. GaS also has a slight gap of about 2.48 eV, but its gap is indirect, which means that the edge of the conduction band and the finesse are not at one point in the space of the wave vector. In this paper, using calculations based on density functional theory, an attempt has been made to investigate the diode properties of GaS nanowires. For this purpose, the parameters related to the calculations ,such as the cut-off energy and the inverting space meshing, are first optimized, and then the GaS unit cell is optimized based on these parameters. Then, the optimal lattice vector and optimal interconnect length were used to continue the calculations. Then, using the density of states calculations, the electronic properties of the GaS transport channel with different impurities were investigated. Then, by calculating the current-voltage diagram, the electronic transport of the GaS transport channel with impurities of different atoms was calculated and investigated.

2. The Theoretical Method

Quantum mechanics has improved our understanding of the structure and behavior of atoms, molecules, solids, and subatomic particles. Although classical physics is well adapted to explaining occurrences in the macroscopic realm, the study of all elements of the subatomic cosmos necessitates instruments that allow for more extensive and complicated computations. Density function theory may be used to calculate the quantum states of atoms, molecules, solids, and molecular dynamics. The one-electron theory is an approach used in computational density physics. This theory gives a method for solving the Schrodinger equation of a very particle-rich system and determining its ground state.

The one-electron theory is an approach used in computational density physics. This theory gives a method for solving the Schrodinger equation of a very particle-rich system and determining its ground state and ground state energy [21].

2.1 Density functional theory

The single electron theory includes the Born-Oppenheimer and Hartry-Fock approximations. Following the definition of electron density and the Thomas-Fermi model, the Schrodinger equation converts a many-particle system into a single-particle equation and is one of the most critical methods. What exists in this theory is the density functional theory (DFT). This method is one of the most efficient computational methods for many-particle systems ,and this is due to its high computational efficiency, reduction in computational volume ,and good accuracy compared to other methods. The importance of DFT becomes even more apparent when we know that Cohen, one of the founders of the method, was awarded the Nobel Prize in Chemistry in 1998. Cohen and Sham considered a hypothetical device consisting of electrons that have no interaction, and eventually established equations that they used to calculate the ground-state properties of the device self-consistently. In density functional theory, the total energy is a combination of three sets: one is the fraction of the kinetic energy of the electrons, one is the fraction of the Coulomb energy due to the electrostatic interaction of all charged particles in the system, and the third is the exchange-correlation energy, which includes all interactions. The system consists of many particles. Among these sets, the exchange correlation energy fraction has an unknown form. It is not possible to find this energy in general. Therefore, approximation methods such as local density approximation (LDA) and generalized gradient approximation (GGA) should be used. The flowchart of self-consistent solution steps in DFT calculations is shown in Figure 1.

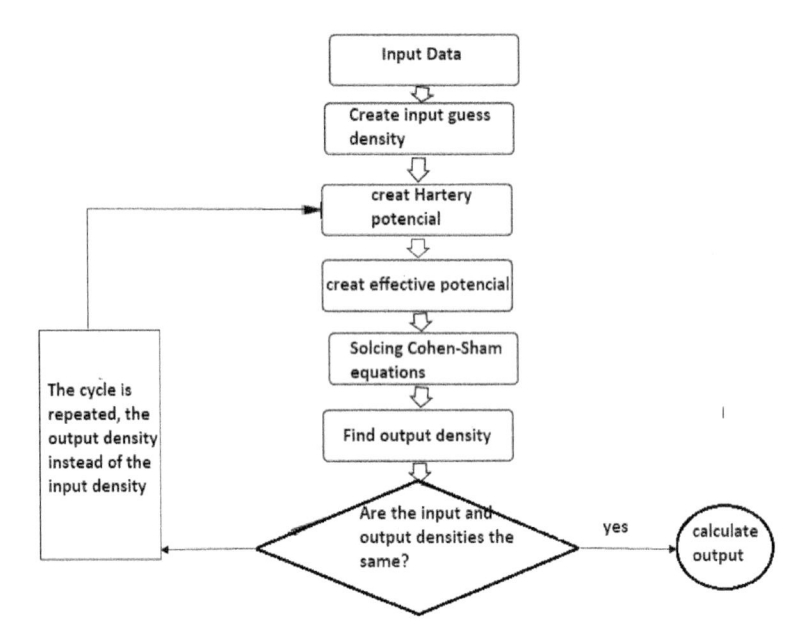

Figure 1. DFT calculations flow chart

Density functional theory deals with an n-electron system whose wave function is a variable n3-functional system. In this theory, all features of the electronic structure of the system, including the interaction of electrons in the external potential (potentials caused by nuclei), are determined by the density of the electron charge. Note that it is only a function of three variables, which reduces the computational volume. Density functional theory transforms the Schrodinger equation into a set of one-particle systems into a set of one-particle equations. These one-particle equations are known as Cohen-Sham equations, which depend only on the electron density [21].

The one-particle problem forms the study of atoms, molecules, solids, gases, liquids, etc. consisting of electrons and nuclei. Quantum mechanics can describe an interaction system of electrons and nuclei by the Schrodinger equation:

$$\hat{H}\Psi = E\Psi \quad (1)$$

Where E is the specific energy value, and the multiarticulate Hamiltonian operator ($\hat{H}$) is given by Equation 2:

$$\hat{H} = \sum_{i=1}^{N=} -\hbar^2/2m\, \nabla_i^2 + \sum_{i=1}^{N\,Nare} - \hbar^2/2M\, \nabla_t^2 + \frac{1}{2}\sum_{i-j} e^2/|r_i - r_j| - \sum_{i,s} Z_i e2/|r_i - R_i| + \frac{1}{2}\sum_{\gamma=1} Z_i Z_j e2/|R_r - R|$$

(2)

In equation (2), the first and second parts are the kinetic energies of electrons and nuclei, and the last three parts are the interactions between electron-electron, electron-nucleus, and nucleus-nucleus, respectively. Planck's constant, m and M are the masses of electron and nucleus respectively, z_I is the atomic number of the 1st atom, e is the charge of the electron, r_i and R_I are the position of the 1st electron and the 1st nucleus.

Solving the Schrodinger equation is a very complex particle. To simplify the problem, the Born-Oppenheimer approximation is used, which states that nuclei are much heavier than electrons and move much slower than electrons. As a result, the motion of nuclei and electrons can be separated. It is assumed that the position of the atoms is fixed while the electrons move in the charge field of the nuclei. The general wave function can be divided into electronic and ionic wave functions. As a result, the Schrodinger equation for the electron part is given by Equation 3:

$$\hat{H}_e(r.R)\Psi_e = E_e \Psi_e(r.R) \quad (3)$$

The Hamiltonian electron operator is also expressed by Equation 4 below:

$$\hat{H}_e = \sum_{i=1}^{N_e} -\hbar^2/2m \, \nabla_i^2 + \frac{1}{2}\sum_{i \neq j} e^2 / |r_i - r_j| + \hat{V}_{ext} \quad (4)$$

In this equation; it is the potential that the nuclei exert on the electrons. Note that the interaction between the nuclei is entered as a parameter. Although the number of degrees of freedom of the system can be reduced by the Born-Oppenheimer approximation, the electron-electron interaction problem is still challenging to solve. Moreover, the electron wave function depends on the coordinates of all electrons. The number of electrons is much higher than the number of nuclei. As we will show below, it is more practical to use the density functional theory instead of the multi-particle wave function. Density functional theory requires fewer calculations and provides a good description of the electronic properties of the ground state of the system.

In density functional theory, the calculations are solved self-consistently. The computational steps of density functional theory can be expressed as follows: 1- Input data: The input data are the coordinates, the number of atoms ,and the total number of electrons. When the quasi-potential is used, we need to know its clear form, and when flat waves are used, we need to look at how the results depend on some critical parameters. For example, the number of points for the Brillouin domain. 2- Input density: In this step, we set the initial experimental density, which can be the overlap of atomic charge densities or comes from quasi-experimental calculations. 3- Generation of the Harter potential: here, we use the Equation of place ,which relates the second potential derivative to the charge density. 4- Generation of exchange potential: the exchange potential is calculated for the density entered. 5- Generation of the effective potential: this potential is obtained by adding the three components of the external exchange potential (V_ext), the Harter potential (V_Hartree) ,and the exchange-correlation potential (V_xc), and we show it with V_effect. 6- Solving the Cohen-Sham equations: At this stage, the Cohen-Sham equations are solved in both direct space and inverse space. 7- Determining the output density: by solving the Cohen-Sham equations, the output density is determined. 8- Testing the degree of convergence: In this phase, the difference between the input density and the output density is calculated. If it is above a specific value, which is set in advance, the cycle must be repeated. However, if the difference is not significant, we have obtained the density of the ground-state.

3. Results and discussion

In this paper, an attempt is made to investigate the diode properties of GaS nanowires using calculations based on density functional theory. For this purpose, the parameters related to the calculations, such as the cut-off energy and the inverting space meshing, are first optimized, and then the GaS unit cell is optimized based on these parameters. Then, the optimal grating vector and interconnect length were used to continue the calculations.

Before performing the corresponding basic calculations and presenting their results, several parameters should be considered to increase the accuracy, decrease the computational cost, and increase the convergence speed. These parameters include the energy for cutting and meshing the inverted space.

3.1 Cut off energy

In computational codes, Cohen Sham single-particle wave functions are expanded according to flat base waves. Since the sentences of this expansion are gradually reduced for high-energy flat waves, these sentences can be considered zero for wave vectors whose energy exceeds a limit value called the cut-off energy. Moreover kept only the sentences whose kinetic energy is lower than the cut-off energy according to Equation (5). This limits the number of base functions in the expansion and reduces the computational volume:

$$\hbar^2/2m \, |\vec{k} + \vec{G}|^2 \leq E_{cut} \qquad (5)$$

K and G are the wave functions and the coordinates of other symmetrical points relative to this point will be as follows respectively:

$$G = \left(\frac{2\pi}{\sqrt{3}a}, 0\right)$$

$$K = \left(\frac{2\pi}{\sqrt{3}a}, 0\right)$$

To obtain the optimal value of this parameter, all parameters except the cut-off energy were selected in an accurate and large input file to ensure that those parameters themselves do not cause errors, then change the cut-off energy at each step and Total energy was evaluated for different cut-off energies. The accuracy of energy calculation in the input files up to 0.001 eV was considered. Perdew-Burke-Ernzerhof (PBE) was used to increase the accuracy of the results of the exchange-correlation function.

The results (Figure 2) showed that the changes in total energy in terms of cut-off energy for the ,NC approximation after 600 eV have a uniform trend and energy changes are less than 0.001 eV. Therefore, this value was selected as the optimal parameter for the cut-off energy in the whole calculation process.

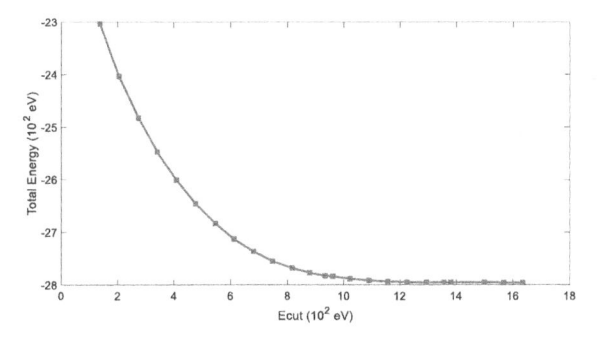

Figure 2. Graph of total energy in terms of cut-off energy (eV) for a GaS structure with quasi-potential NC.

3.2 Grid vector and number of K points in inverted space

In an infinite periodic solid, the number of atoms and electrons is huge. Since single-particle wave functions are spread over all solids, to calculate quantities such as charge density n (r) at the point r, a large number of wave functions that are finely distributed in space must be calculated. To be. Using Bloch's theorem, we can reduce the computational problem Ne ~(10^{23}) the number of single-particle wave functions ψ_i (r $\vec{}$) to the problem ψ_nk (r $\vec{}$) in the first Brillouin region for a limited number of bands.

$$\psi_n(\vec{k}, \vec{r} + \vec{R}) = \psi_n(\vec{k}, \vec{r}) e^{(\vec{\imath k}.\vec{R})} \qquad (6)$$

Since the wave functions of the points k $\vec{}$ that are close together are very similar, the integration on all points k can be approximated by summing on a discontinuous set of points k. Although we need to know the wave functions for all k-points in the first Brillouin region, in practice it is sufficient to have wave functions in a limited number of these points. To integrate these points, we need to network the inverted space. The method in quantum espresso code is the Menkharest-Pack method [22]. Networking is done in the freedom and periodic directions of the structure, which is generally N × M × K.

These parameters are numbers (integers, N, M, K ≠0) to network the inverted space in different directions. For systems with three degrees of freedom ,such as heap systems (N × M × K), for two-dimensional networks (such as graphene) with two degrees of freedom in the direction a $\vec{}$ and b $\vec{}$ and the degree of restriction in the direction c $\vec{}$ (N × N × 1), for one-dimensional and quasi-one-dimensional structures such as nanotubes, nanowires ,and nanowires with a degree of freedom in the direction of c $\vec{}$ as 1 × 1 × N and for structures with three degrees of restriction in the direction of three a $\vec{}$, b $\vec{}$, c $\vec{}$, like molecules, are performed as 1 × 1 × 1, with point Γ at the center of the Brillouin region. The larger the cell and structure we are studying in real space, the smaller its inverted (Brillouin area) will be, and vice versa. Therefore, the smaller the inverted space, the fewer meshes ,and k points in the Brillouin area we need.

To obtain the optimal value of this parameter, all parameters except the inverted space mesh were selected in an accurate and large input file to ensure that those parameters themselves do not cause errors. The results of the previous step are used for the cut-off energy. Then the inverted space meshing is changed in each step, and the total energy for different inverted space meshing is investigated and reported in Figure (3).

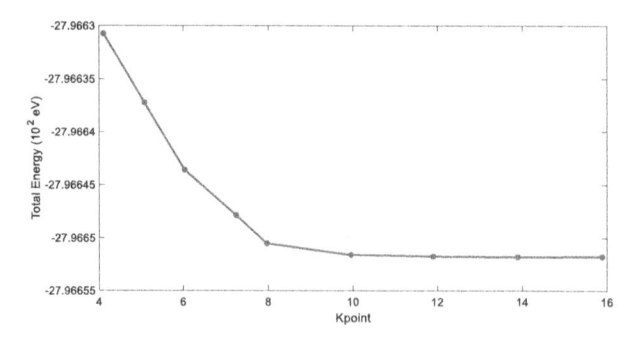

Figure 3. Total energy diagram in inverse space meshing for GaS structure with quasi-potential NC.

According to the obtained results (Figure 3), the changes of total energy in terms of inverse space meshing for NC approximation after meshing 8 × 1 8 1, have a uniform trend, and energy changes are less than 0.001 eV. Therefore, 8×1×1 mesh was selected as the optimal parameter in all calculations.

3.3 Dimensional optimization of structures

After optimizing the parameters required to perform the calculations, the first action to be performed before examining the mechanical, electronic, and optical properties is to optimize the structure and position of the atoms. In this section, ionic dynamics and interatomic forces are studied by the Molecular Dynamics (MD) approach. There are different molecular dynamics methods for this, the BFGS method was used in the calculations, by which the atomic positions and positions are constantly changing, and in different states of force between them by the Hellman-Feynman method until The energy of the whole structure and system continues to be minimized for the position of atoms and ions. By optimizing the structure, atomic positions, equilibrium lattice constants, and the length of atomic bonds can be achieved in equilibrium.

The unit cell of a single-layer GaS structure has four atoms. The cubic unit cells of these structures are shown in Figure (4).

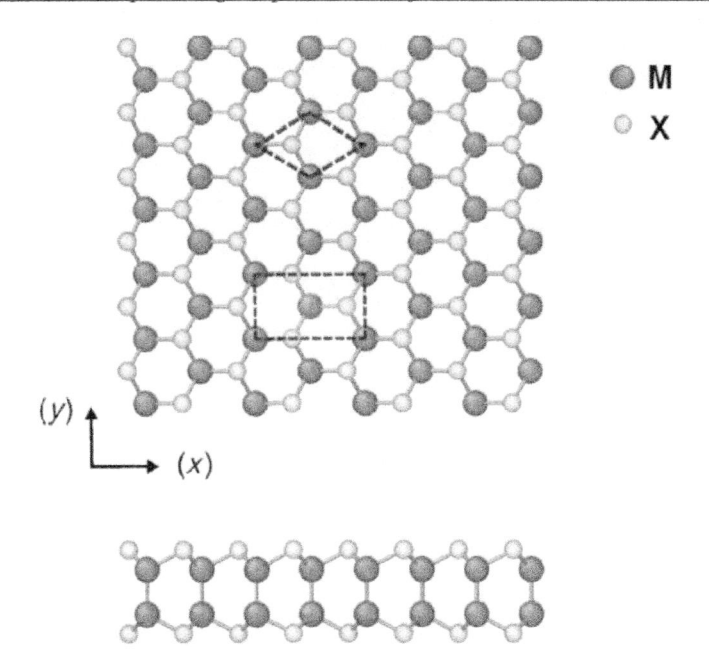

Figure 4. The unit cell has a two-dimensional GaS structure from top to bottom

After dimensional optimization, the new lattice vectors obtained for the GaS structure are reported in Table 1 and compared with previous theoretical results. The values calculated in this work are marked with an asterisk in the table.

case	Approximate	a (Å)	ref
GaS	DFT and LDA and GGA	3.43	This work
	LDA	3.58	[23]
	LDA	3.53	[24]
	GGA	3.48	[25]
	GGA	3.4	[26]

3.4 Electronic transport calculations of GaS structure

In the first step, the electronic properties and transport properties of two-dimensional GaS nanostructures have been studied. For this purpose, a section of the GaS two-dimensional structure is connected to two positive and negative electrodes (Figure 5). The electrodes are also made of GaS.

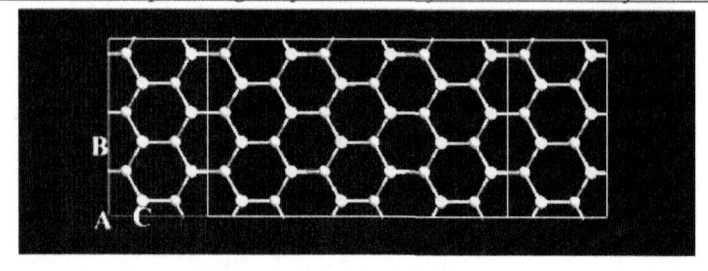

Figure 5. GaS transport channel structure and right and left electrodes connected to it made of GaS.

Using the DFT approach, a diagram of the density of the electronic states of the GaS transport channel is plotted (Figure 6). From this diagram, we find that the bandgap is a structure of 2.2 eV, which is consistent with the previous results ([23-27]). In addition, according to the diagram in Figure 6, it can be seen that the band gap in the GaS structure is not symmetric around the Fermi level and the Fermi level is closer to the edge of the capacity band, so the load carriers in this structure can be holes.

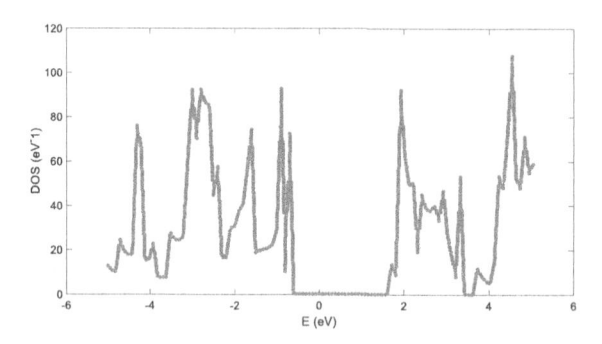

Figure 6. Electronic state density (DOS) diagram of GaS structure.

After examining the electronic properties of the GaS structure, the current-voltage diagram (I-V) related to this structure was calculated (Figure 7). At this stage, the flow through this GaS was investigated using the non-equilibrium Green function approach. This diagram is plotted in -5 to 5 + B bias to show GaS behavior in both positive and negative biases. According to Figure 7, it can be seen that among the voltages of -4 to +4, the behavior of GaS in positive and negative biases is similar ,and for more voltages of 2.2 V, it passes through the current transmission channel. The passing voltage (2.2 V) is the same as the 2.2 eV bandgap of the GaS structure, after which the valence electrons of the valence can be excited and pass through the bandgap to reach the conduction band and in electrical transmission.

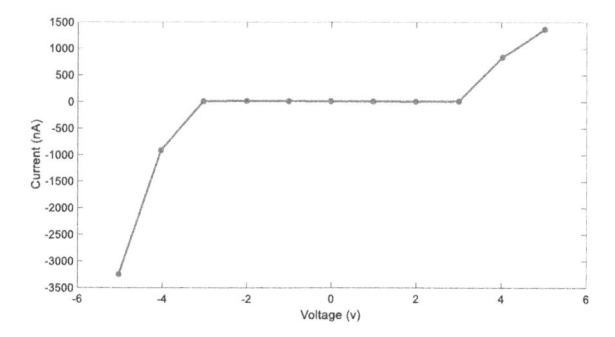

Figure 7. Current-voltage diagram of GaS structure at -5 V to 5V bias.

3.5 Defects by impurities in the structure of GaS

In order to investigate the effect of different defects or impurities (n-type impurities or p-type impurities) on the electronic properties of the GaS structure, we substituted atoms II, IV, V ,and VII instead of Ga and S atoms. Thus, an attempt has been made to design a diode based on a GaS structure using impurities of type n and p. Figure 8 shows the density of states calculated after the doping of the single atoms Ca, Mg, Ba and Sr instead of the Ga atom to the right of the transport channel. These atoms are group II elements that electronegatively tend to lose their electrons in the crystal lattice of the GaS structure and act as n-type impurities in the structure. Thus, by increasing the number of load carriers in the crystal lattice of the GaS transmission channel, they lead to faster and easier transmission of electrical current in the transmission channel. According to the diagram in Figure 8, after the doping of group II atoms instead of the Ga atom to the right of the GaS transport channel, the GaS semiconductor material has become a conductive material. Because after the doping of the single atoms Ca, Mg, Ba ,and Sr, peaks appear on the Fermi level in the state density diagram, which indicates the conduction of GaS. In this way, electrons can be quickly excited from the valence band under the influence of external voltage and transferred to the conduction band.

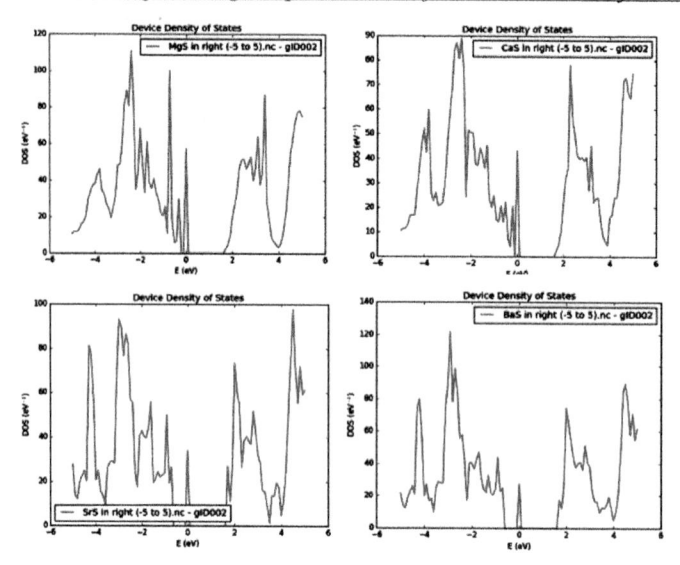

Figure 8. The density of states related to the structure of GaS after the doping of the single atoms Ca, Mg, Ba, and Sr instead of the Ga atom to the right of the transport channel.

In the next step, group IV atoms were used for impurities on the right side of the GaS transport channel instead of the Ga atom (Figure 9). After the doping of the single atoms Ca, Mg, Ba ,and Sr, instead of the Ga atom, as in group II atoms, a peak form is formed. The presence of an electron state on the Fermi level in the state density diagram allows electrons to transition from a valence level to a conduction level. Thus, with the doping of group IV atoms in the GaS transport channel, this material is conductive.

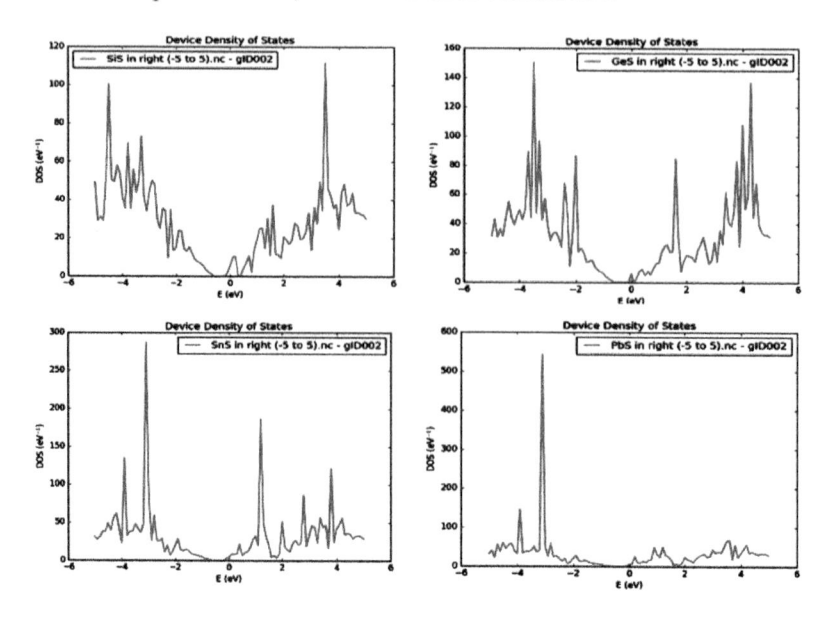

Figure 9. The density of states related to the structure of GaS after the doping of the single atoms Ge, Si, Pb ,and Sn instead of the Ga atom to the right of the transport channel.

In the next step, an attempt is made to change the properties with the impurities of group II atoms (Sr, Mg, and Ca) on the left side of the transport channel and group IV atoms (Sn, Si, and Ge) on the right side of the transport channel instead of Ga atoms. GaS transport channel electrons are calculated. The density of the calculated modes is shown in Figure 10. The calculated state densities show that in addition to the conduction channel of GaS under the influence of impurities of group II and IV atoms, many peaks are formed inside the GaS bandgap, which also increases the probability of band-to-band transitions inside. The light gap has increased. In the previous case, where only one atom of group II elements was doped to one side of the transport channel, the situation differed.

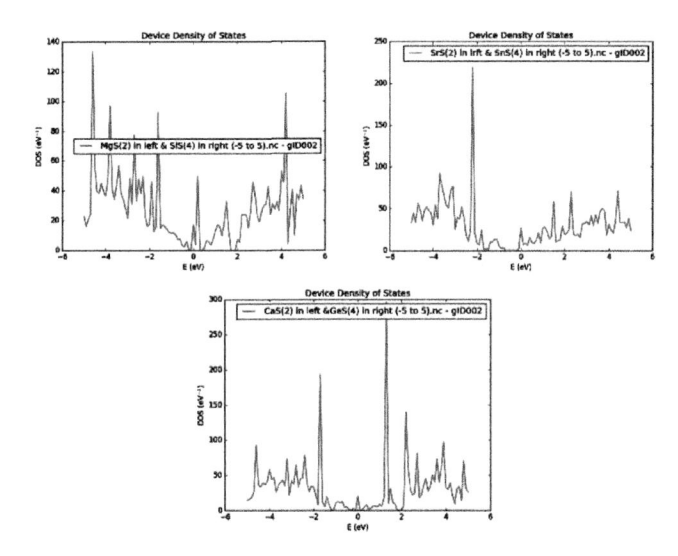

Figure 10. The density of states related to the structure of GaS after the doping of the single atoms Sn, Si, and Ge instead of the Ga atom on the right side of the transport channel and the doping of the Sr, Mg, and Ca atoms instead of the Ga atom on the left side of the transport channel.

In Figure 11, the density states of the GaS structure are calculated and plotted after the impurities of the single atoms Cl, Br, At, and I instead of the S atom to the right of the transport channel. With the doping of group VII atoms instead of S atoms in the GaS transport channel, states are created on the form level that leads to the conduction of the GaS structure.

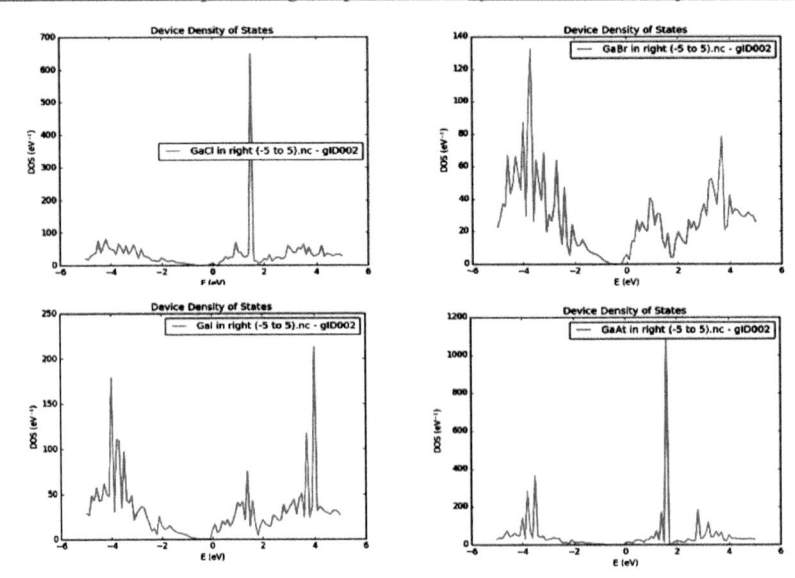

Figure 11. The density of states related to the structure of GaS after the doping of the single atoms Cl, Br, At ,and I instead of the S atom to the right of the transport channel.

In Figure 12, the density states of the GaS structure are plotted after the impurities of the single atoms Br, Cl, and At instead of the S atom to the right of the transport channel, and the impurities of the At, P, and Bi atoms instead of the S atom to the left of the transmission channel. Thus, the effect of impurities of group VII and V elements on the right and left of the transport channel on the electronic properties of the GaS transport channel was studied. According to the density diagram, the states drawn on the Fermi level have been created ,and these structures have also become conductive.

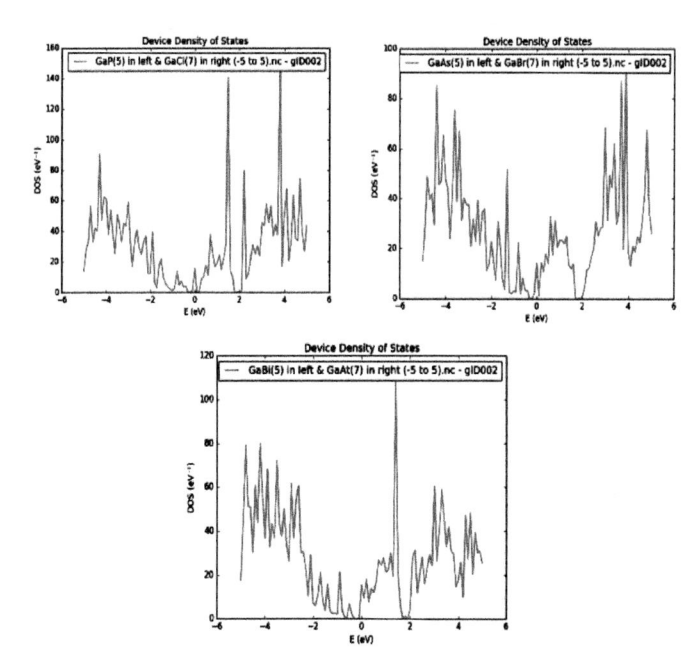

Figure 12. The density of states related to the structure of GaS after the doping of the single atoms Br, Cl, and At instead of the S atom on the right side of the transport channel and the doping of the As, P ,and Bi atoms instead of the S atom on the left side of the transport channel.

Then, in order to investigate the transport properties and the behavior of the GaS transport channel diode, a current-voltage diagram is drawn in the bias of -5 to +5 V. Figure 13 shows the current-voltage diagram of the GaS structure after the doping of the single atoms Ca, Mg, Ba and Sr instead of the Ga atom to the right of the transmission channel. What can be seen in the current-voltage diagram is that at positive bias, up to +3 V, the current was zero. In all these structures, after the voltage of +3 V, the current has gradually increased. The only difference is that the Mg atom is doped instead of the Ga atom. In this case, the trend of increasing current at +4 V suddenly decreases at +5 V from 7 μA to 2.5 μA. The maximum amount of current belongs to the Mg doping state, which at a voltage of +4, a current of 7 μA passes through the transmission channel. In reverse bias, in the two states of impurities Ba and Sr instead of the Ga atom, after a voltage of 2 V, the current gradually increases ,and the current intensity is almost equal to the direct bias mode. However, in the case of Ca ,and Mg impurities, the situation is different ,and the phenomenon of negative differential resistance (NDR) has occurred. In both cases, as soon as the reverse bias starts, the current is established in the transmission channel. But after 1V, the current starts to decrease and then increases again at 3V. Thus, the Ba and Sr atoms have little effect on the behavior of the transport channel for current rectification, but the Mg and Ca atoms can show NDR at negative bias.

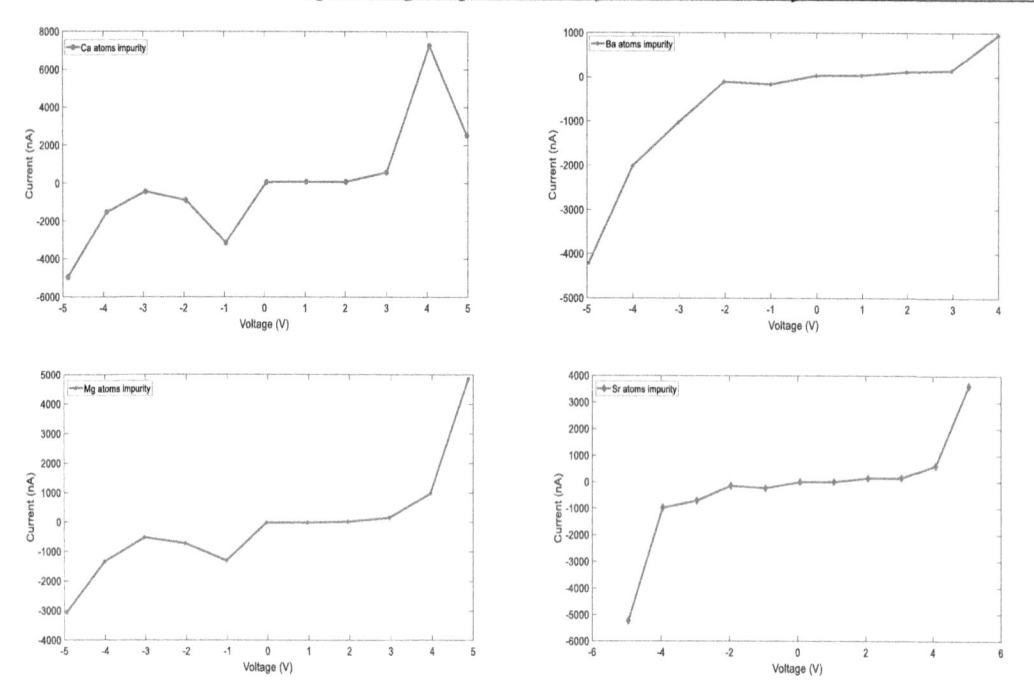

Figure 13. Current-voltage diagram of the GaS structure after doping single Ca, Mg, Ba ,and Sr atoms instead of Ga atoms to the right of the transport channel.

Current-voltage diagram of a GaS structure after doping of single atoms Sn, Si ,and Ge instead of Ga atom on the right side of the transmission channel and doping of Sr, Mg ,and Ca atoms instead of Ga atom on the left side of the transmission channel in Figure 14 given. Calculations show that indirect bias the current intensity is very low and in all three cases ,at +3 V NDR phenomenon has occurred. However, the current intensity indirect bias is not comparable to the current in reverse bias ,and more current in reverse bias is passing through the transmission channel. From the obtained diagrams, it is pretty clear that this type of doping has been able to have a significant effect on the smoothing characteristics of the transport channel. The impurities of the Sn, Si ,and Ge atoms, which belong to group IV, play the role of electron acceptor (p-type) to the system instead of the Ga atom to the right of the transport channel; And the impurities of the Sr, Mg, and Ca atoms instead of the Ga atom on the left of the transport channel play the role of the electron (n-type) in the transport channel. Thus, by creating an n-p type connection, the designed diode could smooth the current passing through the transmission channel. It should also be noted that with the doping of the Mg atom on the left and the Sr atom on the right of the transmission channel, the highest current in reverse bias passed through the transmission channel and was the best rectifier in this category. Most of the current passing through this transmission channel occurred in -5 V reverse bias with an intensity of 30 μA.

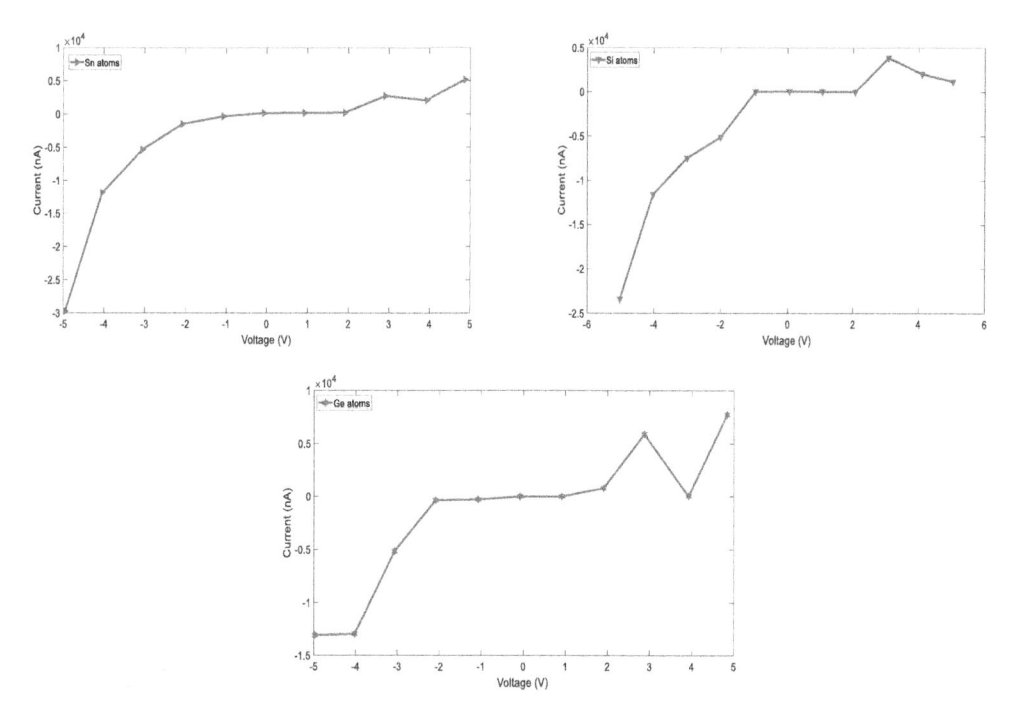

Figure 14. Current-voltage diagram of a GaS structure after doping of single Sn, Si ,and Ge atoms instead of Ga atoms on the right side of the transmission channel and impurities of Sr, Mg ,and Ca atoms instead of Ga atoms on the left side of the transmission channel

The impurities of the single atoms Ge, Si, Pb ,and Sn instead of the Ga atom on the right side of the transport channel could not lead to a definite rectification in the transport channel. In this type of pollution, indirect bias, after a voltage of +1 V, current flows in the system, and after a voltage of 2 to 3 V, the current decreases ,and the NDR phenomenon occurs. At a voltage of +4 V, the current is completely cut off ,and then, with a gradual increase in voltage, the current is restored in the system. In contrast, in reverse bias, after a voltage of 2 V, the current is established and has increased in a quasi-ohmic manner. The only thing that has happened in the reverse bias of the NDR phenomenon is that the Ge atom is doped instead of the Ga atom. In this case, the NDR phenomenon has occurred at -3 to 4 V. Among these structures, most of the current flow was in reverse bias after the doping of the Si atom instead of Ga, which reached more than 16 μA at 5V.

4. Conclusions

In this paper, using calculations based on modified density functional theory, an attempt has been made to investigate the diode properties of GaS nanowires. For this purpose first, by optimizing the parameters related to the calculations such as cut-off energy and inverting space meshing and then using these parameters, the GaS unit cell is optimized. Then the optimal grid vector and the optimal link length were used in to continue the calculations. Then, using state density calculations, the electronic properties of GaS transport channels with different impurities were investigated. Then, by calculating the current-voltage diagram, the electronic transport of GaS transport channel with impurities of different atoms was calculated and investigated.

The DFT approach calculates the electronic properties and transport characteristics of a pure and contaminated GaS transport channel. The graph of the density of electron states related to the GaS transmission channel showed that the bandgap of the structure is 2.2 eV. In addition, the bandgap in the GaS structure is not symmetric around the Fermi level ,and the Fermi level is closer to the edge of the capacity band, so the load carriers in this structure can be cavities. Then, by plotting the current-voltage diagram, it was observed that among the voltages of -4 to +4, the behavior of GaS in positive and negative biases is similar ,and for more voltages, 2.2 V passes through the current transmission channel. The passing voltage (2.2 V) is the same as the bandgap 2.2 eV of the GaS structure, after which the valence electrons of the valence can be excited and pass through the bandgap to reach the conduction band and in electrical transmission. Participate. In order to investigate the effect of different impurities (n-type impurities or p-type impurities) on the electronic properties of the GaS structure, atoms of groups II, IV, V ,and VII were substituted instead of Ga and S atoms. Thus, an attempt has been made to design a diode based on a GaS structure using impurities of type n and p. After the impurities of the atoms of groups II, IV, V and VII instead of the Ga and S atoms in the GaS transport channel, the semiconductor material GaS has become a conductive material. Because after the doping of these atoms, peaks appear on the Fermi level in the state density diagram, which indicates the conduction of GaS. Then, in order to investigate the transport properties and the behavior of the GaS transport channel diode, a current-voltage diagram is drawn in the bias of -5 to +5 V. The Ba and Sr atoms have little effect on the behavior of the transport channel for current rectification, but the Mg and Ca atoms can exhibit NDR at negative bias. When the Mg atom is doped instead of the Ga atom, the current increase in current at +4 V suddenly decreases from + 7 μA to 2.5 μA at +5V. The maximum amount of current belongs to the Mg doping state, which at a voltage of +4, a current of 7 μA passes through the transmission channel. The impurities of the Sn, Si ,and Ge atoms, which belong to group IV, play the role of electron acceptor (p-type) to the system instead of the Ga atom to the right of the transport channel; And the impurities of the Sr, Mg, and Ca atoms instead of the Ga atom on the left of the transport channel play the role of the electron (n-type) in the transport channel. Thus, by creating an n-p type connection, the designed diode could smooth the current passing through the transmission channel. It should also be noted that with the doping of the Mg atom on the left and the Sr atom on the right of the transmission channel, the highest current in reverse bias passed through the transmission channel and was the best rectifier in this category. Most of the current passing through this transmission channel occurred in -5 V reverse bias with an intensity of 30 μA.

References

[1] Wang Q H, Kalantar-Zadeh K, Kis A, Coleman J N and Strano M S 2012 Electronics and optoelectronics of two-dimensional transition metal dichalcogenides Nat. Nanotechnol. 7 699

[2] Nayeri M, Moradinasab M and Fathipour M 2018 The transport and optical sensing properties of MoS2, MoSe2, WS2 and WSe2 semiconducting transition metal dichalcogenides Semicond. Sci. Technol. 33 025002

[3] Yan Z, Yoon M and Kumar S 2018 Influence of defects and doping on phonon transport properties of monolayer MoSe2 2D Mater. 5 031008

[4] Li S L, Tsukagoshi K, Orgiu E and Samorì P 2016 Charge transport and mobility engineering in two-dimensional transition metal chalcogenide semiconductors Chem. Soc. Rev. 45 118–51

[5] Nayeri M and Taheri H 2019 The influence of impurity on the electronic properties of WS2 single layer 2019 27th Iranian Conf. on Electrical Engineering (ICEE) (Piscataway, NJ) (IEEE) pp 224–7

[6] Manzeli S, Ovchinnikov D, Pasquier D, Yazyev O V and Kis A 2017 2D transition metal dichalcogenides Nature Reviews Materials 2 17033

[7] Nayeri M, Fathipour M and Goharrizi A Y 2016 The effect of uniaxial strain on the optical properties of monolayer molybdenum disulfide J. Phys. D: Appl. Phys. 49 455103

[8] Trainer D J, Zhang Y, Bobba F, Xi X, Hla S W and Iavarone M 2019 The effects of atomic-scale strain relaxation on the electronic properties of monolayer MoS2 ACS nano 13 8284–91

[9] Nayeri M and Fathipour M 2018 A numerical analysis of electronic and optical properties of the zigzag MoS2 nanoribbon under uniaxialstrain IEEE Trans. Electron Devices 65 1988–94

[10] Novoselov K S, Mishchenko A, Carvalho A and Neto A C 2016 2D materials and van der Waals heterostructures Science 353 aac9439

[11] Yang S, Jiang C and Wei S H 2017 Gas sensing in 2D materials Applied Physics Reviews 4 021304

[12] Jappor H R 2017 Electronic structure of novel GaS/GaSe heterostructures based on GaS and GaSe monolayers Physica. B 524 109–17

[13] Hu P, Wen Z, Wang L, Tan P and Xiao K 2012 Synthesis of few-layer GaSe nanosheets for high performance photodetectors ACS Nano 6 5988–94

[14] Late D J, Liu B, Luo J, Yan A, Matte H R, Grayson M, Rao C N R and Dravid V P 2012 GaS and GaSe ultrathin layer transistors Adv. Mater. 24 3549–54

[15] Lei S, Ge L, Liu Z, Najmaei S, Shi G, You G, Lou J, Vajtai R and Ajayan P M 2013 Synthesis and photoresponse of large GaSe atomic layers Nano Lett. 13 2777–81

[16] Demirci S, Avazlı N, Durgun E and Cahangirov S 2017 Structural and electronic properties of monolayer group III monochalcogenides Phys. Rev. B 95 115409

[17] Ma Y, Dai Y, Guo M, Yu L and Huang B 2013 Tunable electronic and dielectric behavior of GaS and GaSe monolayers Phys. Chem. Chem. Phys. 15 7098–105

[18] Yagmurcukardes M, Senger R T, Peeters F M and Sahin H 2016 Mechanical properties of monolayer GaS and GaSe crystals Phys. Rev. B 94 245407

[19] Chen H, Li Y, Huang L and Li J 2015 Influential electronic and magnetic properties of the galliumsulfide monolayer by substitutional doping J. Phys. Chem. C 119 29148–56

[20] Li W and Li J 2015 Piezoelectricity in two-dimensional group-III monochalcogenides Nano Research 8 3796–802

[21] Y. Zhou, S. Li, W. Zhou, X. Zu, and F. Gao, "Evidencing the existence of intrinsic half-metallicity and ferromagnetism in zigzag gallium sulfide nanoribbons," Scientific reports, vol. 4, p. 5773, 2014.

[22] H. J. Monkhorst and J. D. Pack, "Special points for Brillouin-zone integrations," Physical review B, vol. 13, no. 12, p. 5188, 1976.

[23] B. P. Bahuguna, L. K. Saini, R. O. Sharma, and B. Tiwari, "Hybrid functional calculations of

electronic and thermoelectric properties of GaS, GaSe, and GaTe monolayers," Physical Chemistry Chemical Physics, vol. 20, no. 45, pp. 28575-28582, 2018.

[24] T. Chen et al., "Ultrathin All-2D Lateral Graphene/GaS/Graphene UV Photodetectors by Direct CVD Growth," ACS Applied Materials & Interfaces, 2019.

[25] B. Zhou et al., "A type-II GaSe/GeS heterobilayer with strain enhanced photovoltaic properties and external electric field effects," Journal of Materials Chemistry C, vol. 8, no. 1, pp. 89-97, 2020.

[26] B. Chitara and A. Ya'akobovitz, "Elastic properties and breaking strengths of GaS, GaSe and GaTe nanosheets," Nanoscale, vol. 10, no. 27, pp. 13022-13027, 2018.

[27] A. Fleurence, R. Friedlein, T. Ozaki, H. Kawai, Y. Wang, and Y. Yamada-Takamura, "Experimental evidence for epitaxial silicene on diboride thin films," Physical review letters, vol. 108, no. 24, p. 245501, 2012.

Peculiarities of the formation of the structure of CMCs based on Al$_2$O$_3$ micropowder and SiC nanopowder in the process of electrosintering

A.G. Mamalis[1], E.S. Gevorkyan[2,3], M.M. Prokopiv[4], V.P. Nerubatskyi[2*], D.A. Hordiienko[2], D. Morozow[3], O.V. Kharchenko[4]

[1]*Project Center for Nanotechnology and Advanced Engineering, Athens, Greece*
[2]*Ukrainian State University of Railway Transport, Kharkiv, Ukraine*
[3]*Kazimierz Pulaski University of Technology and Humanities in Radom, Poland*
[4]*V. Bakul Institute for Superhard Materials of the National Academy of Sciences of Ukraine, Kyiv, Ukraine*

The influence of silicon carbide nanoadditives on the microstructure of the composite based on aluminum oxide microparticles during hot pressing due to electrosintering is investigated. The introduction of silicon carbide nanoparticles into the material leads to the formation of a specific structure of the material. A dispersion-strengthened nanocomposite of the "micro/nano" type is formed. On the other hand, all ceramic material containing granular silicon carbide powder is composite. In the process of sintering, a liquid phase is formed at a lower temperature, compared to the sintering of "micro-micro" particles. Such a complex structure of the composite leads to a significant improvement in the physical and mechanical properties of the material. In addition to the size of the SiC particles, the defects of the crystal lattice of the starting materials have a significant impact on the sintering and properties of the studied compositions. Nanodispersed silicon carbide powder has a more defective crystal lattice compared to the lattice of silicon carbide micropowder. The silicon carbide nanopowders used are more defective and, therefore, more active for sintering, since the process takes place under maximally non-equilibrium conditions. The microstructure obtained with different compositions of nanoadditives and sintering modes is studied. The peculiarities of the effect of various nanoadditives from Al$_2$O$_3$ and SiC on the microstructure and properties of the obtained materials were revealed. It was established that the use of this method makes it possible to improve such mechanical properties of the material as microhardness ($HV = 25.0$ GPa) and crack resistance (KIC = 6.5 MPa·m$^{1/2}$).

Keywords: composite material, nanopowder, electrosintering, liquid phase, crack resistance, nanocomposite, ceramic matrix composite.

1. Introduction

Composite material based on Al_2O_3–SiC powders, due to its unique physical, mechanical and operational properties, has increasingly attracted many scientists specializing in materials science in recent years. Cheap and available on the market starting components made this material competitive compared to other known ceramic materials of this type. CMCs-composites Al_2O_3–SiC are promising as an instrumental and structural material. The creation of functionally gradient materials based on them will make it possible to further expand their prospective use [1–5]. The main trend in the development of modern composite materials, in particular instrumental materials, is to reduce material costs, respect ecology, and increase productivity.

CMCs based on aluminum oxide are interesting as tool materials with the greatest inertness to iron and are used mainly for processing carbon steels [6, 7]. Their main drawback is low strength and crack resistance [8]. The problem was solved by adding other compounds, in particular up to 20% TiC, TiN, TiCN and creating heterophase dispersion-strengthened ceramics by traditional hot pressing at 1600...1650 °C, and this is a high temperature – punches burn with the release of CO_2 [9, 10]. These materials provided clean and semi-clean turning, but did not provide a tendency to increase productivity due to an increase in cutting speed, because the temperature in the cutting zone increased due to insufficient thermal conductivity and strength of the cutting plates. To solve this problem, silicon carbide fibers can be added to aluminum oxide, which significantly increases the properties [11, 12].

Obtaining ceramic matrix composites based on aluminum oxide and silicon carbide by the method of electroconsolidation at temperatures at which chemical interaction occurs between the components with the formation of new phases both in the solid and in the liquid state makes it possible to obtain the maximum level of properties by various methods of hot pressing (electrosintering) [13, 14]. However, in this case, the technology of preservation of fibers during mixing during cold and hot pressing is somewhat complicated. An increase in the temperature of hot pressing leads to an increase in the cost of the material and deterioration of ecology [15, 16].

The use of silicon carbide nanopowder improves the mechanical properties of the composite, and also makes it possible to lower the temperature of hot pressing of the composite material [17]. At the same time, the traditional principles of materials science regarding the formation of the structure and properties of the material are preserved during hot pressing only for temperatures of 1400 °C and 1500 °C. At a temperature of 1600 °C, an increase in polytypes 6-H and IV (Si_7C_7) due to a decrease in the initial cubic polytype 3C was found in the material structure. A slight decrease in hardness and crack resistance was established. When the temperature of hot pressing is increased to 1700 °C, chemical interaction occurs between the components and the complete transformation of atomic 3C into 6H cells of silicon carbide and the formation of up to 4 wt.% SiO_2.

Despite the presence of the liquid SiO_2 phase ($T_p = 1713$ °C), the density of the material decreased by 5% compared to the density of samples obtained at temperatures of 1400 °C and 1500 °C. Disadvantages include the high temperature of the low-productivity method of hot pressing for obtaining dense materials. One of the methods for solving this problem is the use of zirconium oxide in a stoichiometric ratio with Al_2O_3, which corresponds to the eutectic composition [18–20].

Today, the most successful material is based on Al_2O_3 micropowder and up to 15% micro- or nano-SiC [21–23]. At the same time, in the case of using micropowders, regardless of the parameters of the hot pressing process, the structure and properties of the material are formed on the basis of traditional materials science: the phase composition and their ratio correspond to the parameters of the original powder mixture. In the ceramic matrix composite CMCs, obtained on the basis of micropowders and nanopowders of Al_2O_3–SiC by the method of electroconsolidation, interesting transformations occur with the formation of a liquid phase and the disclosure of the features of the formation of the structure of this composite. This makes it possible to improve physical and mechanical properties, especially for the creation of promising functional gradient materials [24, 25].

2. Experimental conditions

The research used micropowders (1...2 μm) of α-Al_2O_3 produced by LLC "Khimlaboreaktiv" (Kyiv, Ukraine), as well as SiC nanopowder of the brand #4629HW manufactured by Nanostructured & Amorphous Materials (USA) CAS #409–21–2 (Fig. 1).

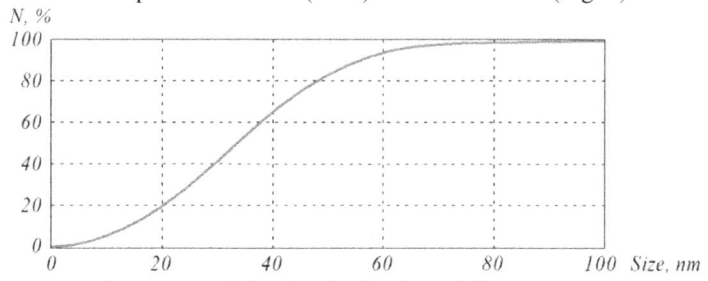

Figure 1. Particle size distribution of SiC100 powder

Basic information on the chemical composition of SiC nanopowder is given in Table 1.

X-ray phase analysis of these powders showed a clearly defined 3C-modification (cubic) phase and the presence of a small amount of 6H-modification (hexagonal) phase (Fig. 2, Fig. 3).

The results of analyzes of nano-SiC type powder show that the average size of SiC crystals is about 100 nm. The predominant phase of SiC is cubic 3C. Analysis of X-ray patterns and Raman scattering spectra indicates the presence of small amounts of the phase of the hexagonal polytype 6H-SiC and carbon.

Based on nano-SiC powder and nano (submicron) Al_2O_3 powder, charges were prepared for further compaction due to electroconsolidation.

Table 1. Chemical composition of SiC nanopowder

Parameter	Value
Color	gray-green
Crystalline form	cubic (3C)
Particle size, nm	80...100
Specific surface area, m²/g	20...30
SiC content, %	99
Si, %	69.1

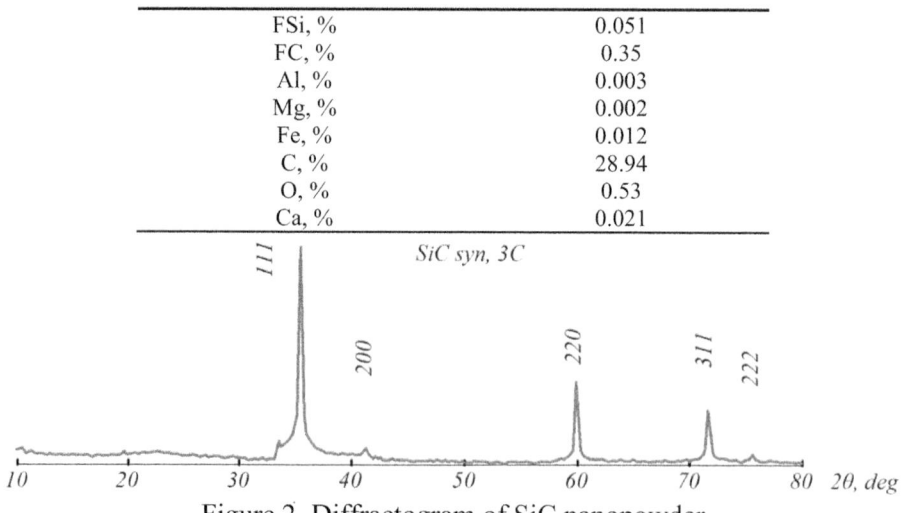

FSi, %	0.051
FC, %	0.35
Al, %	0.003
Mg, %	0.002
Fe, %	0.012
C, %	28.94
O, %	0.53
Ca, %	0.021

Figure 2. Diffractogram of SiC nanopowder

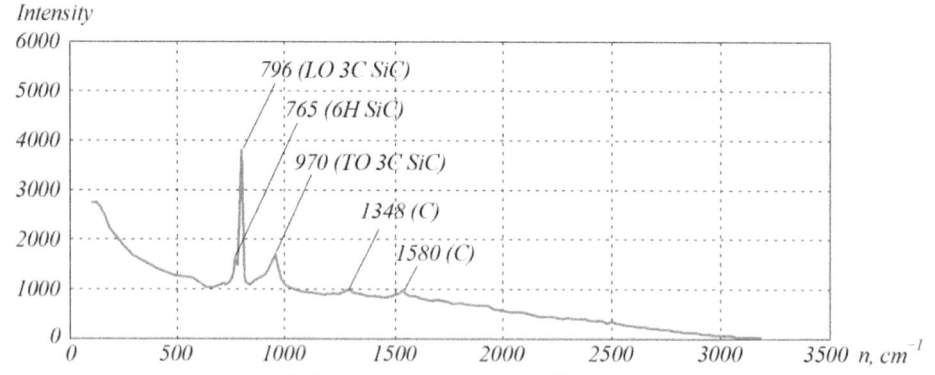

Figure 3. Raman spectrum of SiC nanopowder

According to the results of X-ray phase analysis, the charge consists of a mixture of Al_2O_3 with parameters $a = 4.760$ Å; $c = 12.993$ Å and SiC-3C with a lattice parameter $a = 4.359$ Å (Table 2).

The size of the coherent scattering region practically coincides and is 82.7 nm and 87.7 nm, respectively. Both Al_2O_3 and SiC are in a nanostructured state. The size of the coherent scattering region is 201.5 nm and 68.6 nm, respectively.

Table 2. Parameters of Al_2O_3–SiC powder mixtures

№ mixture	Phase	wt.%	Lattice parameters, Å	D, nm	ε
1	Al_2O_3	85.4	$a = 4.760$; $c = 12.991$	201.5	$3.4 \cdot 10^{-4}$
2	SiC-6H	14.6	$a = 3.083$; $c = 15.110$	68.6	$1.85 \cdot 10^{-3}$

On the diffractograms in Fig. 4, there are additional lines from petrolatum, which is used to bind powders, but the width of the lines is not affected.

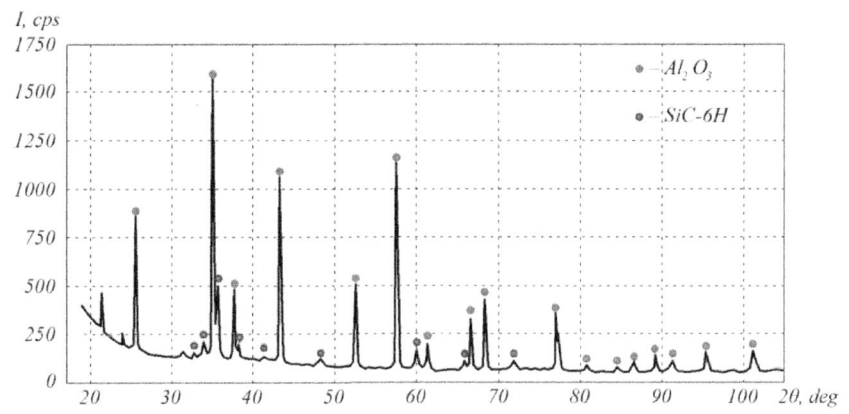

Figure 4. Diffractograms of powders with composition Al₂O₃–15 wt.% SiC

Mixing of the starting powders was carried out in a planetary mill "SAND" with a rotation frequency of 150 rpm in a medium of methyl alcohol for 2 hours. The duration of mixing was determined by the degree of obtaining a homogeneous distribution of powders in the mixture.

Samples for research were obtained in a hot pressing installation, which uses hot pressing by the electroconsolidation method [26, 27] at temperatures of 1400 °C, 1500 °C, 1600 °C, 1700 °C, as well as at $T = 1760$ °C.

Crack resistance was determined on the polished samples according to the Palqvist method [28, 29] under a load of $P = 550$ N on the Vickers pyramid. The hardness was determined by Rockwell on the TK-2 device and by Vickers under a load on the diamond pyramid $P = 300$ N. Microhardness was measured on the PMT3 device under a load of 2 N with simultaneous fixation of the microstructure and an impression from the Vickers pyramid. The chemical integral composition was determined using scanning electron microscope. X-ray microspectral studies of the local and integral composition of the structure of the sample obtained at 1760 °C were carried out using a scanning electron microscope "Carl Zeiss" (Germany).

3. Results and discussion

In Fig. 5 shows the surface structures of the samples in the optical image, which were obtained at temperatures of 1400 °C and 1500 °C. It should be noted that in the optical image of the structure of this material, the dark phase is aluminum oxide, and the light phase is silicon carbide.

A   B

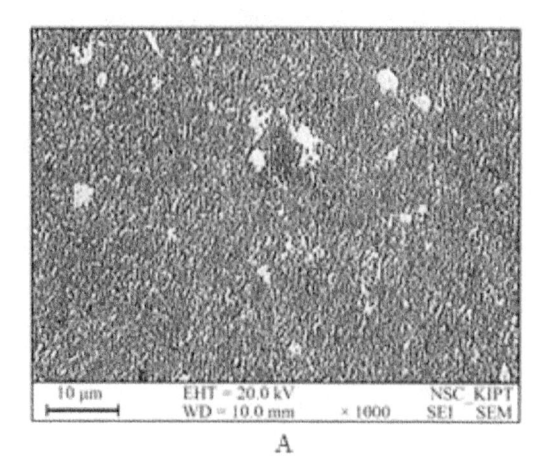

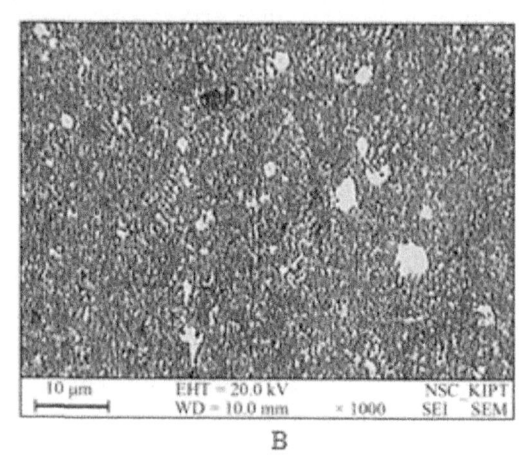

Figure 5. Image of the structure of the surface of the sample obtained at temperatures of 1400 °C, 1500 °C with exposure time of 3 min: (A) view of the structure of the extreme part of the surface of the cut sample, which is made at an angle of 2...3 degrees; (B) the middle of the sample with impressions from the Vickers pyramid

As can be seen from Fig. 5, the unprocessed surface of the sample has a characteristic uneven "bumpy" appearance, as after hot pressing, which makes it difficult to determine its structure. After diamond treatment removes the irregularities of this surface at an angle of 2...3° to a depth of 10^{-5} µm, it can be seen that the main structure (Figure 5, b) includes a dark-colored matrix phase with a small content of stochastically distributed dispersed inclusions 1...5 µm in size light phase. Individual grains of the light phase up to 20 µm in size were also found in the structure. As research has shown, the thickness of this surface layer is 30...50 µm. The hardness of *HV* is 18.7 GPa. The main structure also has a matrix phase, but lighter than the previous color, in which there are also dispersed inclusions of a light phase similar in size and distribution, but with a much smaller content.

At sintering temperatures exceeding 1800 °C, areas with a secondary eutectic microstructure formed from the liquid phase are found in the microstructure of the pressings (Fig. 6), which indicates locally inhomogeneous heating of the pressing material under the conditions of spark-plasma sintering. Analyzing visually the photographs of the structures of the samples obtained in the temperature range of 1400...1700 °C, we can say that they are of the same type in structure. The dark phase is the matrix, and the light phase is the dispersed particles of silicon carbide. Moreover, the size is in the range of 1...20 µm. That is, in the process of hot pressing, nanopowder of silicon carbide is formed in an aggregated state of light inclusions. There is a slight difference in the structure of the sample obtained at a temperature of 1760 °C.

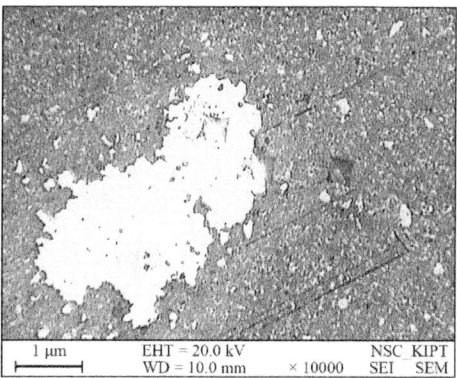

Figure 6. The microstructure obtained from a mixture of Al2O3 (micro)–15 wt.% SiC (nano) at a temperature of 1800 °C, a pressure of 35 MPa and a holding time of 3 min

In the temperature range of 1400...1700 °C, the structure in the center of the sample does not change and is a matrix of the dark-colored phase.

On the surface of the sample in Fig. 7, spots with a complex structural structure, as well as a light phase with dispersed inclusions of a dark and light gray phase, were detected.

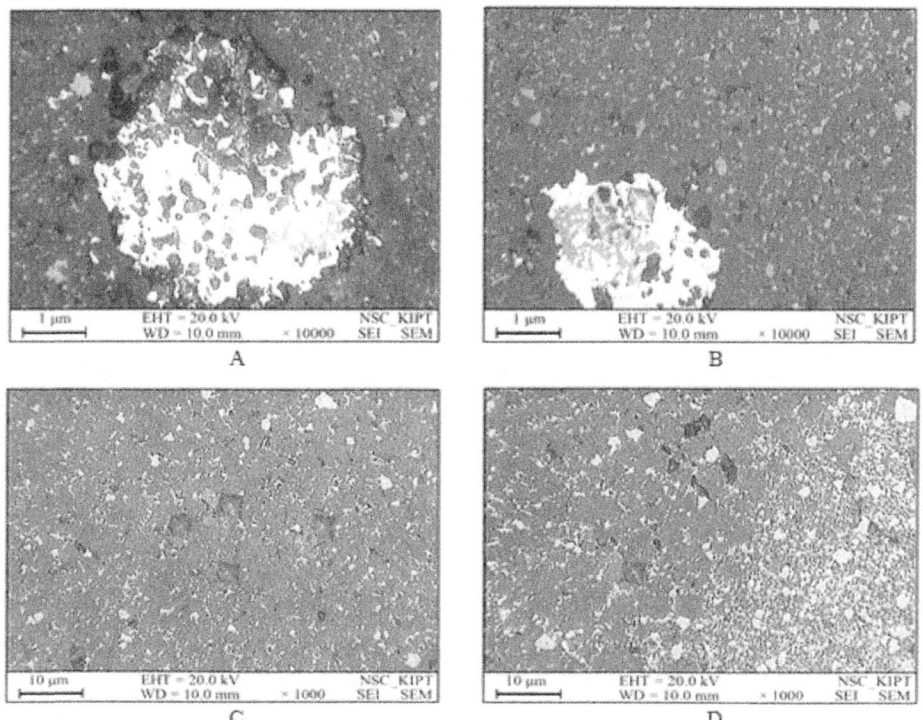

Figure 7. Structures of different parts of the surface of the sample obtained at a temperature of 1760 °C from a mixture of Al2O3 (micro)–15 wt.% SiC (nano): (A) and (B) aggregates of

SiC nanopowders; (C) and (D) hardness meter prints in different parts of the sample
In Fig. 8 shows that as a result of the formation of a liquid phase in the composite, silicon carbide particles are arranged in the form of mesh colonies.

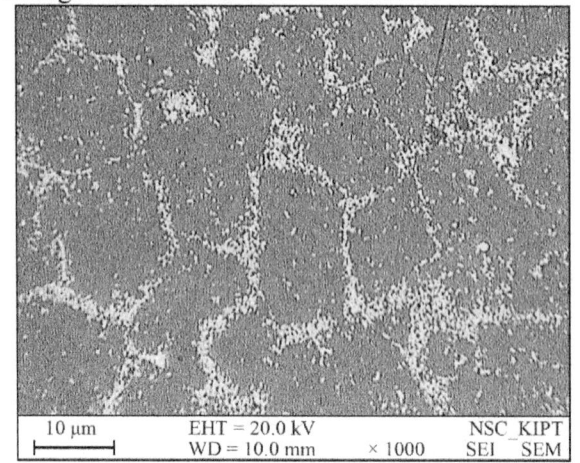

Figure 8. The surface structure of the sample obtained at a temperature of 1760 °C from a mixture of Al$_2$O$_3$ (micro)–15 wt.% SiC (nano)

In Table 3 shows the results of the integral chemical analysis of the elements in the structure of the samples depending on the sintering temperature.
The structure of the Al$_2$O$_3$ (micro)–15 wt.% SiC sample is shown in Fig. 9, and the micro X-ray spectral analysis of the sample is shown in Table 4. From the Table 4 shows that free carbon appears in the structures obtained at a temperature of 1700 °C.

Table 3. Integral chemical analysis of elements in the structure of samples

Sintering temperature, °C	C, %	O, %	Al, %	Si, %	Fe, %	W, %
1400	7.20	44.62	37.81	9.92	0.22	0.12
1500	6.92	44.81	37.69	9.85	0.66	0.06
1600	6.57	46.51	39.21	8.22	0.38	0.00
1760	3.31	47.85	40.92	6.23	0.06	1.63

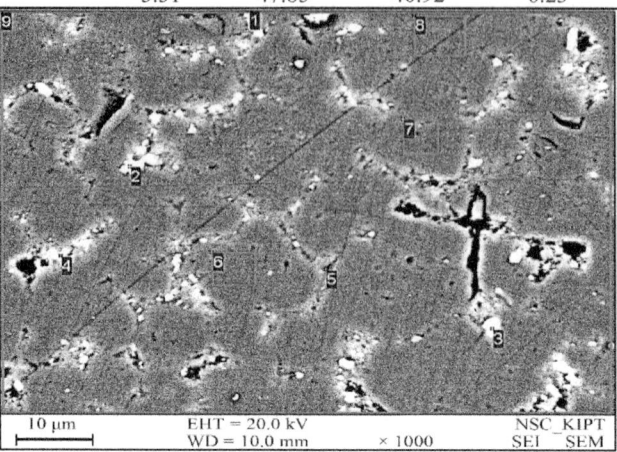

Figure 9. Structures of different parts of the surface of the sample obtained at a temperature of 1700 °C from a mixture of Al$_2$O$_3$ (micro)–15 wt.% SiC (nano)

Table 4. Results of chemical analysis of elements in the structure

Spectrum	C, %	O, %	Al, %	Si, %	Fe, %	Co, %	W, %
1	61.57	13.23	7.26	17.57	–	–	0.37
2	51.99	19.51	11.00	16.08	0.44	0.31	0.66
3	51.01	18.98	11.85	16.40	0.18	0.75	0.86
4	59.88	8.60	12.80	18.05	–	0.15	1.51
5	27.68	37.32	18.60	16.39	–	–	–
6	10.13	58.02	30.58	1.28	–	–	–
7	8.28	59.03	30.06	2.62	–	–	–
8	8.77	59.67	30.96	0.59	–	–	–
9	13.39	53.87	26.99	5.74	–	–	–

The structure obtained from a mixture of Al2O3 (micro)–15 wt.% SiC (nano) is shown in Fig. 10.
In the Table 5 shows the X-ray microspectral analysis of the sample area in squares 9 and 10.

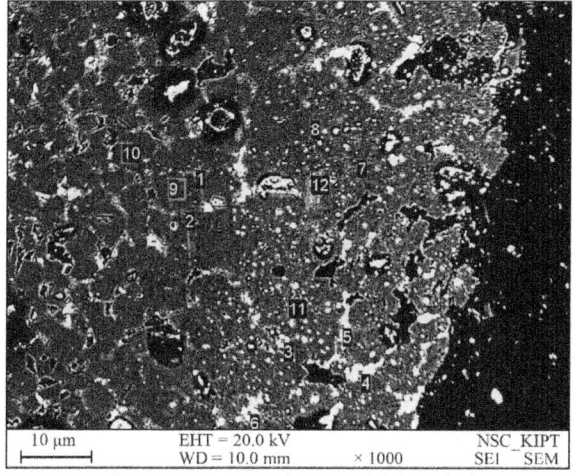

Figure 10. The structure obtained from a mixture of Al_2O_3 (micro)–15 wt.% SiC (nano)

Table 5. X-ray microspectral analysis of the sample area in squares 9 and 10

Spectrum	C, %	O, %	Al, %	Si, %	Ti, %	Co, %	W, %
1	27.78	38.08	19.33	14.80	–	–	–
2	13.53	52.62	29.93	3.93	–	–	–
3	58.04	–	3.49	37.77	–	–	0.71
4	59.14	2.40	3.57	34.16	0.73	–	–
5	41.73	26.47	9.57	21.81	0.13	–	0.29
6	62.57	6.55	7.41	22.05	0.22	–	1.20
7	29.34	44.29	20.36	6.02	–	–	–
8	19.93	47.67	24.38	8.02	–	–	–
9	8.88	59.31	30.09	1.73	–	–	–
10	9.37	58.49	30.61	1.53	–	–	–
11	24.40	43.06	22.83	9.72	–	–	–
12	39.16	27.44	14.24	18.97	–	0.18	0.01

In the case of using SiC nanopowder, the temperature of hot pressing of the material with maximum properties is decreases by 350...400 °C, while the traditional principles of materials science of the formation of the structure and properties of the material are

preserved during hot pressing only for temperatures of 1400 °C and 1500 °C.
At a temperature of 1600 °C, an increase in polytypes 6H and IV (Si$_7$C$_7$) due to a decrease in the initial cubic polytype 3C was found in the material structure (Fig. 11).
A slight decrease in hardness and crack resistance was established. Increasing the temperature of hot pressing to 1700 °C between the components leads to chemical interaction and complete transformation of the atomic 3C to 6H cell of silicon carbide and the formation of up to 3.26 wt.% SiO$_2$. Although it is generally known that the transition temperature is above 2000 °C. The reduction of aluminum oxide and silicon carbide directly indicates that silicon oxide was formed as a result of the interaction between them: Al$_2$O$_3$ – 82.85 wt.%; SiC-6H – 13.28 wt.%; SiO$_2$-stishovite – 3.26 wt.%; SiO$_2$-Quartz – 0.61 wt.%.

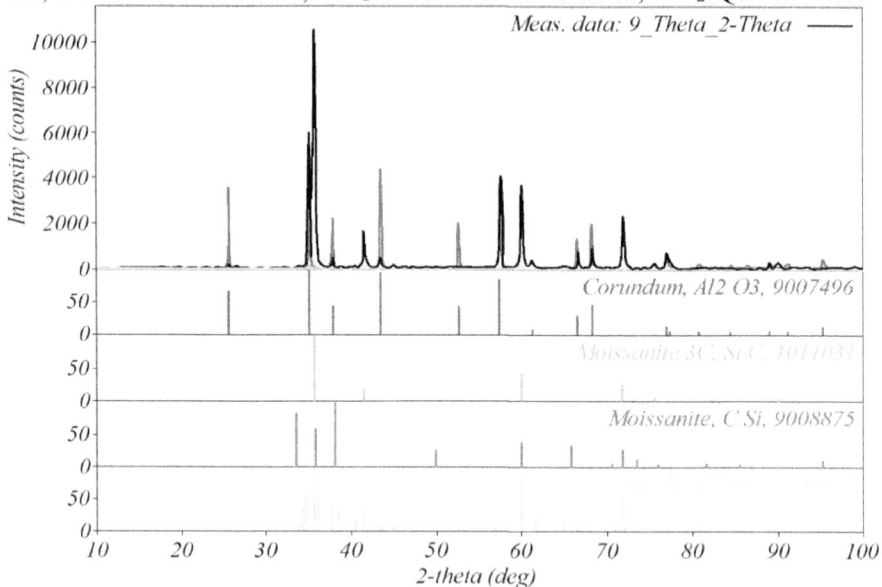

Figure 11. X-ray phase analysis of Al$_2$O$_3$–15 wt.% SiC composites obtained by hot pressing due to electrosintering at temperature T = 1600 °C, pressure P = 35 MPa, sintering time 2 min

Today, well-known materials are obtained from both macro- and micropowders of the original components by hot pressing at a temperature of T = 1800 °C, friction pairs have been created [30], as well as cutting plates for high-performance mechanical processing of alloys on based on iron. In the case of using micropowders during hot pressing at temperatures up to 1800 °C, a classic matrix-type structure is formed from Al$_2$O$_3$ with a SiC dispersed phase and the corresponding formation of a complex of properties (Table 6).

Table 6. Results of measurements of hardness and crack resistance of the Al$_2$O$_3$ sample

Sample temperature	Crack resistance K_{IC}, MPa·m$^{1/2}$	Hardness HV (15)
1500 °C	5.3	13.5 (edge)
1500 °C	5.1	14.0 (between them)
1500 °C	6,1	15.6 (center)
1400 °C	4.1	17.9 (edge)
1400 °C	5,3	18.8 (center)
1700 °C	4.1	16.0 (center)
1700 °C	4.5	14.6 (edge)

The research was carried out with loads on the Vickers pyramid – 15 kg. Some observations during the preparation of grinding samples on the ShP-3A planetary grinding and polishing machine.

At 1760 °C, a more uniform distribution of the light gray phase is observed, as well as a deformation from a round to an elongated shape. A sample made from a mixture of Al_2O_3 micro-15 wt.% (Fig. 12).

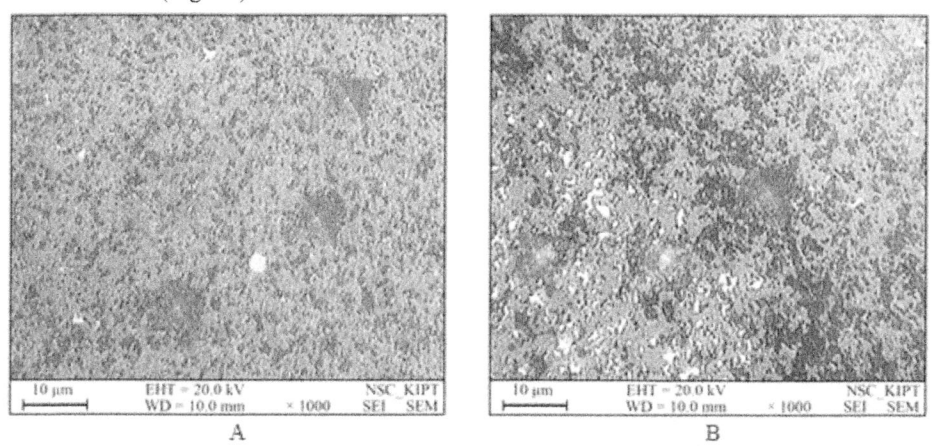

Figure 12. The structure obtained from a mixture of Al_2O_3 (micro)–15 wt.% SiC (nano) at a temperature of 1760 °C

As can be seen, compared to the initial composition, the mass fraction has changed, which is caused by chemical reactions. Thus, the SiO_2 phase is formed in the composition of the sample after consolidation (Fig. 13).

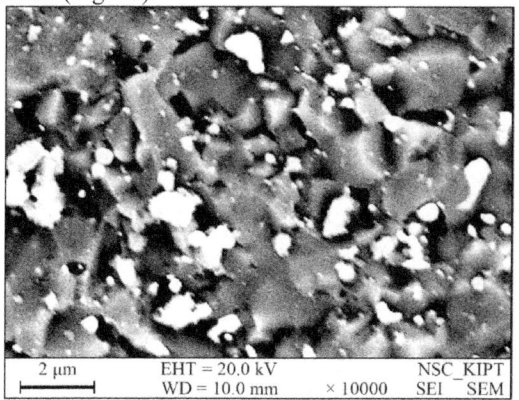

Figure 13. Microstructure of Al_2O_3–15 wt.% SiC composites

In Fig. 14 shows the dependence of the change in relative density on temperature at different electrosintering temperatures.

The advantage of using nanodisperse powders in liquid-phase sintering is to lower the sintering temperature while maintaining the necessary properties. An increase in the dispersion of the powder increases the defectivity of its crystal lattice and the reserve of excess energy.

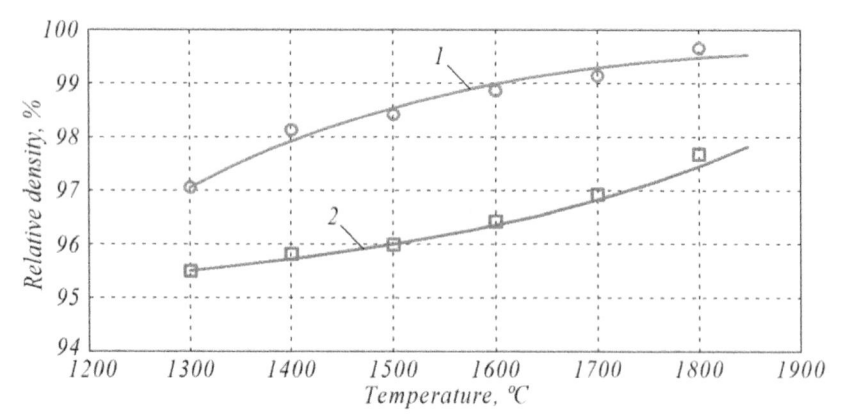

Figure 14. Relative densites CMCs Al$_2$O$_3$–15 wt.% SiC under different temperature during electrosintering: *1* – SiC (nano); *2* – SiC (micro)

An increase in the surface layer with distorted lattices promotes the flow of surface diffusion and mass transfer. This approach refers to physical methods of sintering activation.

Grain growth can be controlled by reducing the sintering temperature and holding time, but this causes poor sintering of the particles due to incomplete grain boundary diffusion. As the content of SiC particles in the initial mixture increases, the average grain size decreases. The presence of SiC nanoadditive particles prevents the growth of grains, without requiring a decrease in the sintering temperature and holding time.

4. Conclusions

Obtaining composite materials with high mechanical properties during the consolidation of powder materials is associated with solving a number of difficult tasks. On the one hand, an increase in the sintering temperature promotes a more active course of grain boundary diffusion and their sintering, on the other hand, it activates the formation of by-products, which negatively affects the density and hardness of the material.

When choosing the optimal compaction parameters of composite materials from nanosized powders, special attention should also be paid to their increased activity. But on the other hand, this feature of nanopowders can have a positive effect during liquid-phase sintering, lowering the temperature of eutectic mixture formation.

The study of the mechanical properties of the obtained composites showed that the Al$_2$O$_3$–SiC composite reaches the maximum values of microhardness (HV = 25.0 GPa) and crack resistance (K_{IC} = 6.5 MPa·m$^{1/2}$) with a composition of 85 wt.% micro-Al$_2$O$_3$–15 wt.% nano-SiC, sintering temperature T_s = 1600 °C, exposure time t = 2 min.

Funding: The researches were co-financed by Ministry of Education and Science of Ukraine project No. 0121U109441.

References

1. M. Cao, Y. Liu, S. Sun, H. Zhou, X. Meng, C. Wang, Effect of Al2O3 micro-powder size on mechanical properties and mi-crostructure of Al2O3-SiC composite ceramic. Ceramics International 41 (2015) 5533–5539. DOI: 10.1016/j.ceramint.2014.12.048.

2. S.L. Cai, X.M. Lu, X.L. Wang, X.C. Hu, Microstructure and mechanical properties of Al2O3/SiC nanocomposite ceramics. Journal of Alloys and Compounds 662 (2016) 85–91. DOI: 10.1016/j.jallcom.2015.12.209.

3. W. Li, W. Li, Q. Li, Y. Liu, Z. Zhang, H. Li, Effect of SiC nanopowder on properties and microstructure of Al2O3-SiC composite ceramic. Ceramics International 44 (2018) 9652–9657. DOI: 10.1016/j.ceramint.2018.02.197.

4. A. Sharma, A. Thakur, R.K. Dutta, Effect of SiC reinforcement on Al2O3 matrix by powder metallurgy technique. Materials Today: Proceedings 5 (2018) 11313–11319. DOI: 10.1016/j.matpr.2018.03.157.

5. S.K. Padhi, S.C. Panigrahi, S.K. Sahoo, Mechanical and tribological behaviour of Al2O3-SiC composite: A review. Tribology - Materials, Surfaces & Interfaces 14 (2020) 97–105. DOI: 10.1080/17515831.2020.1778767.

6. G. Karadimas, K. Salonitis, Ceramic matrix composites for aero engine applications—a review. Applied Sciences 13 (2023) 3017. DOI: 10.3390/app13053017.

7. S. Marimuthu, J. Dunleavey, Y. Liu, M. Antar, B. Smith, Laser cutting of aluminium-alumina metal matrix composite. Optics & Laser Technology 117 (2019) 251–259. DOI: 10.1016/j.optlastec.2019.04.029.

8. C. Bach, F. Wehner, J. Sieder-Katzmann, Investigations on an all-oxide ceramic composites based on Al2O3 fibres and alumina–zirconia matrix for application in liquid rocket engines. Aerospace 9 (2022) 684. DOI: 10.3390/aerospace9110684.

9. P. Putyra, M. Podsiadlo, B. Smuk, Alumina-Ti(C,N) ceramics with TiB2 additives. Archives of Materials Science and Engineering 47/1 (2011) 27–32.

10. L.A. Dobrzanski, M. Kremzer, A. Nagel, B. Huchler, Fabrication of ceramic preforms based on Al2O3 CL 2500 powder. Journal of Achievements in Materials and Manufacturing Engineering 18 (2006) 71–74.

11. E. Gevorkyan, A. Mamalis, R. Vovk, Z. Semiatkowski, D. Morozow, V. Nerubatskyi, O. Morozova, Special features of manufacturing cutting inserts from nanocomposite material Al2O3-SiC. Journal of Instrumentation 16, 10 (2021) P10015. DOI: 10.1088/1748-0221/16/10/P10015.

12. E.S. Gevorkyan, V.P. Nerubatskyi, R.V. Vovk, O.M. Morozova, V.O. Chyshkala, Yu.G. Gutsalenko, Revealing thermomechanical properties of Al2O3–C–SiC composites at sintering. Functional Materials 29, 2 (2022) 193–201. DOI: 10.15407/fm29.02.193.

13. Z. Krzysiak, E. Gevorkyan, V. Nerubatskyi, M. Rucki, V. Chyshkala, J. Caban, T. Mazur, Peculiarities of the phase formation during electroconsolidation of Al2O3–SiO2–ZrO2 powders mixtures. Materials 15, 17 (2022) 6073. DOI: 10.3390/ma15176073.

14. E.S. Gevorkyan, V.P. Nerubatskyi, R.V. Vovk, V.O. Chyshkala, M.V. Kislitsa, Structure formation in silicon carbide – alumina composites during electroconsolidation. Journal of Superhard Materials 44, 5 (2022) 339–349. DOI: 10.3103/S1063457622050033.

15. D. Sofronov, A. Krasnopyorova, N. Efimova, A. Oreshina, E. Bryleva, G. Yuhno, S. Lavrynenko, M. Rucki, Extraction of radionuclides of cerium, europium, cobalt and strontium with Mn3O4, MnO2, and MNOOH sorbents. Process Safety and Environmental Protection 125 (2019) 157–163. DOI: 10.1016/j.psep.2019.03.013.

16. E. Gevorkyan, L. Cepova, M. Rucki, V. Nerubatskyi, D. Morozow, W. Zurowski, V. Barsamyan, K. Kouril, Activated sintering of Cr2O3-based composites by hot pressing. Materials 15, 17 (2022) 5960. DOI: 10.3390/ma15175960.

17. V.P. Nerubatskyi, R.V. Vovk, M. Gzik-Szumiata, E.S. Gevorkyan, Investigation of the effect of

silicon carbide nanoadditives on the structure and properties of microfine corundum during electroconsolidation. Fizika Nizkikh Temperatur 49, 4 (2023) 540–546.

18. J. Merisalu, T. Jogiaas, T.D. Viskus, A. Kasikov, P. Ritslaid, T. Kaambre, A. Tarre, J. Kozlova, H. Mandar, A. Tamm, Structure and electrical properties of zirconium-aluminum-oxide films engineered by atomic layer deposition. Coatings 12 (2022) 431. DOI: 10.3390/coatings12040431.

19. E. Gevorkyan, V. Nerubatskyi, Yu. Gutsalenko, O. Melnik, L. Voloshyna, Examination of patterns in obtaining porous structures from submicron aluminum oxide powder and its mixtures. Eastern-European Journal of Enterprise Technologies 6, 6(108) (2020) 41–49. DOI: 10.15587/1729-4061.2020.216733.

20. E. Gevorkyan, V. Nerubatskyi, V. Chyshkala, O. Morozova, Revealing specific features of structure formation in composites based on nanopowders of synthesized zirconium dioxide. Eastern-European Journal of Enterprise Technologies 5, 12(113) (2021) 6–19. DOI: 10.15587/1729-4061.2021.242503.

21. E.S. Gevorkyan, M. Rucki, A.A. Kagramanyan, V.P. Nerubatskiy, Composite material for instrumental applications based on micro powder Al2O3 with additives nano-powder SiC. International Journal of Refractory Metals and Hard Materials 82 (2019) 336–339. DOI: 10.1016/j.ijrmhm.2019.05.010.

22. M.I. Abd El Aal, H.H. El-Fahhar, A.Y. Mohamed, E.A. Gadallah, The mechanical properties of aluminum metal matrix composites processed by high-pressure torsion and powder metallurgy. Materials 15 (2022) 8827. DOI: 10.3390/ma15248827.

23. K. Raja, V. Sridhar, Mechanical performance of nano-copper reinforced with SiC, AL2O3 using powder metallurgy process. Journal of Emerging Technologies and Innovative Research 6, 5 (2019) 569–574. https://www.jetir.org/papers/JETIRCU06111.pdf.

24. O. Fomin, A. Lovska, V. Pistek, P. Kucera, Research of stability of containers in the combined trains during transportation by railroad ferry. MM Science Journal (2020) 3728–3733. DOI: 10.17973/MMSJ.2020_03_2019043.

25. O. Fomin, A. Lovska, V. Masliyev, A. Tsymbaliuk, O. Burlutski, Determining strength indicators for the bearing structure of a covered wagon's body made from round pipes when transported by a railroad ferry. Eastern-European Journal of Enterprise Technologies 1, 7(97) (2019) 33–40. DOI: 10.15587/1729-4061.2019.154282.

26. E. Gevorkyan, M. Rucki, Z. Krzysiak, V. Chishkala, W. Zurowski, W. Kucharczyk, V. Barsamyan, V. Nerubatskyi, T. Mazur, D. Morozow, Z. Siemiatkowski, J. Caban, Analysis of the electroconsolidation process of fine-dispersed structures out of hot pressed Al2O3–WC nanopowders. Materials 14, 21 (2021) 6503. DOI: 10.3390/ma14216503.

27. E. Gevorkyan, V. Nerubatskyi, V. Chyshkala, Y. Gutsalenko, O. Morozova, Determining the influence of ultra-dispersed aluminum nitride impurities on the structure and physical-mechanical properties of tool ceramics. Eastern-European Journal of Enterprise Technologies 6, 12(114) (2021) 40–52. DOI: 10.15587/1729-4061.2021.245938.

28. V.A. Lapitskaya, T.A. Kuznetsova, A.V. Khabarava, S.A. Chizhik, S.M. Aizikovich, E.V. Sadyrin, B.I. Mitrin, W. Sun, The use of AFM in assessing the crack resistance of silicon wafers of various orientations. Engineering Fracture Mechanics (2021) 107926. DOI: 10.1016/j.engfracmech.2021.107926.

29. D.V. Prosvirnin, M.E. Prutskov, M.D. Larionov, A.G. Kolmakov, Evaluation of the method of measuring crack resistance by the introduction of the vickers indentor for aluminum oxynitride ceramics. Journal of Physics: Conference Series 1431 (2020) 012047. DOI: 10.1088/1742-6596/1431/1/012047.

30. O.P. Umans′kyi, A.H. Dovhal′, A.D. Panasyuk, O.D. Kostenko, Vplyv skladu i struktury kompozytsiynoyi keramiky na osnovi karbidu kremniyu na mekhanizmy znoshuvannya. Poroshkova metalurhiya 07/08 (2012) 92–102 (in Ukrainian).

Thermodynamic Assessment of the Fe–Mn Binary Alloy System using the CALPHAD Method

D.F. Khan[1], W.U. Khan[1], W.U. Shah[1*], H.U. Shah[1], H.Q. Yin[2], A.G.Mamalis[3]

[1]*Department of Physics, University of Science and Technology, Bannu 28100, Khyber Pakhtunkhwa, Pakistan*
[2]*School of Materials Science and Engineering, University of Science and Technology, 100083 Beijing, P.R. China*
[3]*Project Centre for Nanotechnology and Advanced ngineering (PC-NAE), NCSR–Demokritos, Athens, Greece*
Email: waseemullahshah303@gmail.com

Thermodynamic assessment of and predictions for the metastable binary iron–manganese alloy (Fe–Mn system) have been undertaken using the CALPHAD (calculation of phase diagram) method. Phase diagrams, Gibbs energies of mixing, excess Gibbs energies, thermodynamic molar activities, coefficient of activities, and partial and integral values of enthalpy have been calculated at three elevated temperatures: 1200, 1225 and 1250 K. The alloy shows positive deviations from Vegard's and Henry's laws (the latter deviation being small) and corresponding negative deviation from Raoult's law (ideal Gibbs curve). Results show that the ferromagnetic state of the Fe–Mn alloy is the most stable. Phase equilibria show almost ideal characteristic behaviour. Overall the alloy is at equilibrium and has good stability.

Keywords: PBIN database, phase diagram.

1. Introduction
Mn is a very well reported element for the evolution of significant magnetic steels such as TWIP steels, which have good strength and ductility. Lattice stability of Mn and Fe was first investigated using the CALPHAD method by Schwerdtfeger and S. Bigdeli [1-2]. The method has been used for the developing and modeling thermodynamic properties of many materials for four decades. A complete CALPHAD-based assessment of Fe–Mn system was carried out by A.T. Dinsdale and W.Huang. [3-4]. The thermochemical approach for the Fe–Mn system was formulated by B.J. Lee et al [5]. The martensitic-like transformation for

Fe-based alloys regarding the sigma phase (fcc-A1) and Gibbs energy optimization was studied in refs [6] and [7]. Eutectoid-based reactions for the Fe–Mn alloy system appear to be relatively insignificant . Magnetism plays a major role in the stabilization of the gamma phase at the lower eutectoid temperature. The activity of Mn in the alloy was assessed using CALPHAD [8-9].

The present research into the Fe–Mn system uses the third generation of the Thermo-Calc software package and CALPHAD database. The optimization of the Fe–Mn system and the Fe–Mn–Si system transformation enthalpy was assessed.

2. Results and Discussion
2.1 Phase diagram

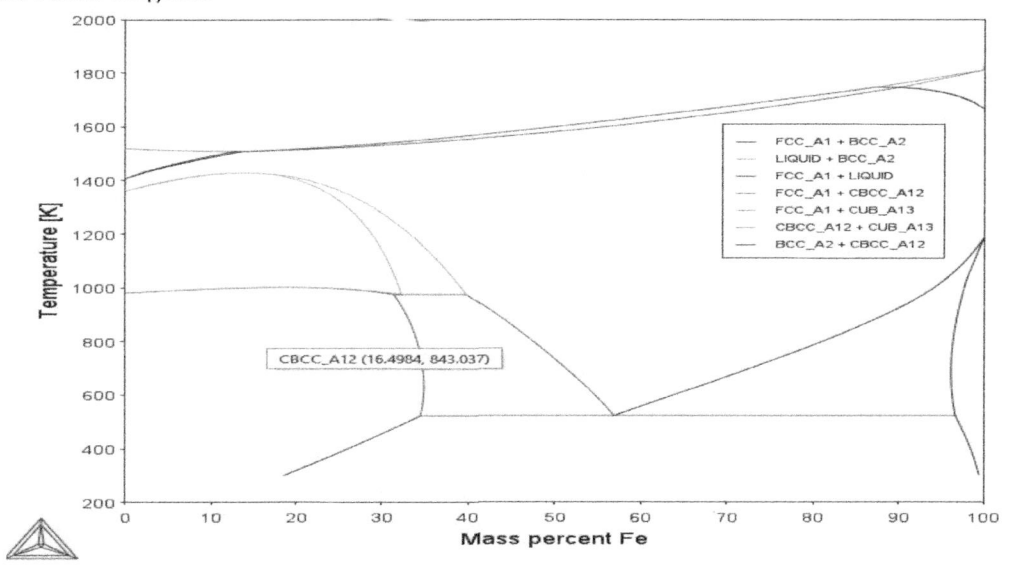

Figure 1. Computed phase diagram of the Fe-Mn alloy, shows face centred cubic, liquid, cumulative bace centered cubic.the abscissa gives x_{Mn}, the fraction of Mn atoms in the alloy.

Fig. 1 shows the phase diagram of the ferric-manganese binary alloy system by the applications of Calphad method with Thermo-Calc package.different natured solid and liquid phases have investigated by applying temperature range (1200-1250K) .(10-90) mass percent of Fe results the highest stability phase FCC with austenite coordinations.CBCC phase with 16.4984 mass% of the Mn results the corresponding stability in connection to BCC regim with ferrite coordination.the alloy seems less metastable range for whole range compositions. Results are in good agreement with available experimental results and previous assessments by E.J. Mittemeijer [10]. The phase transformation FCC-A1 (austenite), BCC-A2 (ferrite phase), and FCC-A1/HCP-A3 (martensitic) of the phase diagram calculated by researchers in the Fe-Mn system with all solution phases as liquid,BCC-A12,BCC-A2,CUB-A13,FCC-A1,CPH-A3, CPH-A3 phases and shows metastable one accordance as the phases have calculated by Redlich-kister polynomial calculations .Mn as a transition metal plays a vital role in the creation of magnetic properties for the said alloy system. The CPH-A3 phase being metastable creates homogenous heterogeneity in the phase structure of the system and makes variational based characteristics for the solid solution. Previous research results show that the ferromagnetic state of Mn-Fe-C alloy is the most stable . Investigated results report many magnetic-based moments of atoms [11]. For Mn and Fe system, their measured phase equilibria show activity curve data for solid in almost ideal behavior, but somehow small positive deviation from ideality . Stable phase diagram for the Fe-Mn system has been investigated in accordance with A. Christou , D.D. Johnson, and D. Birnies recommendations[12-14]. The ferromagnetic phase is the most stable phase in that particular alloy system with occurring at the ground state. As a transition metal alloys the magnetic properties if the said alloy is very complex natured depends on the alternation of thermal energy and curie temperatures. The magnetism is responsible for creating stable nature properties in that alloy system with low-temperature ranges [15]

$$G^q = x_{Fe}{}^0 G_{Fe}{}^q + x_{Mn}{}^0 G_{Mn}{}^q + G_m{}^q + G_{mag}{}^q \qquad (1)$$

Where G is the Gibbs free energy of a system is for pure element, for Mn, Fe, Gibbs energy paramagnetic state value is 298.15k. , while the other parameters x, $G_m{}^q$ molar fraction, molar Gibbs energy, and magnetic-based ordering of the system,

$$G_m{}^q = RT[x_{Fe} \ln x_{Fe} + x_{Mn} \ln x_{Mn}] + G^{qs} \qquad (2)$$

While R is gas constant with excess molar Gibbs energy. Here the Redlich-Kister power series is used for mathematical manipulations as

$$G_N{}^{q,s} = x_{Fe} x_{Mn} \sum_{v=0}^{n} L^q Fe, Mn (x_{Fe} - x_{Mn})^v \qquad (3)$$

Now by Calphad convention rule

$$^v L^q Fe, Mn = a_v{}^q + b_v{}^q T + c_v{}^q T \ln T + d_v{}^q T^2 + e_v{}^q T^3 + f_v{}^q T^{-1} + g_v{}^q T^7 + h_v{}^q T^{-9} \qquad (4)$$

With "a" and "h" are the imperial parameters. $^v L^q$ is the angular momentum term for v-q state. While $a_v{}^q, b_v{}^q, c_v{}^q, d_v{}^q, e_v{}^q, f_v{}^q, g_v{}^q, h_v{}^q$ are the activities of the a-h states in magnetic orbital transition of the particular alloy system. xFe, xMn shows the composition range for both alloys.

Here the bcc-A2 of ferromagnetic phase shows magnetic contribution to this particular system, bcc-A12, cph-A3, fcc-A1, while cph-A3 shows anti ferromagnetic contributions to this system, as in accordance with A.T. Dinsdale calculations [16]

$$G_{mag}{}^q = RT \ln (B^q + 1) g(t) \qquad (5)$$

While for numerical solution, if g (t) <1 then we will observe as

$$g(t) = 1 - (79^{t-1}/140 p + 474/497(1/p - 1)(t^3/6 + t^9/135 + t^{15}/600))/D \qquad (6)$$

$$G(t) = - (t^{-5}/10 + t^{-15}/315 + t^{-25}/1500)/D \quad \text{if} \quad t > 1 \qquad (7)$$

While $D = 0.46044 + 0.73189(1/(p-1)) \qquad (8)$

$t = T/T^q{}_{C,N}$ that represents a particular curie temperature of **q-Th** phase during transformation from ferromagnetic to paramagnetic transformation, similarly anti ferromagnetic to paramagnetic . $G_{mag}{}^q$, shows the magnetic part of gibbs energy of orbitals. RT shows the enternal energy corrwsponding to entropy of the system. information about average magnetic moment, as well the crystal structure of that particular alloy system [17], the value of **P** is 0.40 for bcc-A2,while for fcc-A1,BCC-A12,CPH-A3 are 0.38.so

$$C_{pmag}{}^q = RT \ln (Bq + 1) C(t) \qquad (9)$$
$$C(t) = 474/497(1/p - 1)(2r^3 + 2t^9/3 + 2t^{15}/5)) \qquad (10)$$
$$D, \quad \text{if } t < 1 \quad C(t) = (2t^{-5} + 2t^{-15}/3 + 2t^{-25}/5)/D \quad \text{if } t > 1 \qquad (11)$$

The proposed prediction in the Fe-Mn alloy system shows contribution to the ferromagnetic and anti-ferromagnetic phases and their transformation as in accordance with literature through Calphad method.

3. Thermodynamic modeling

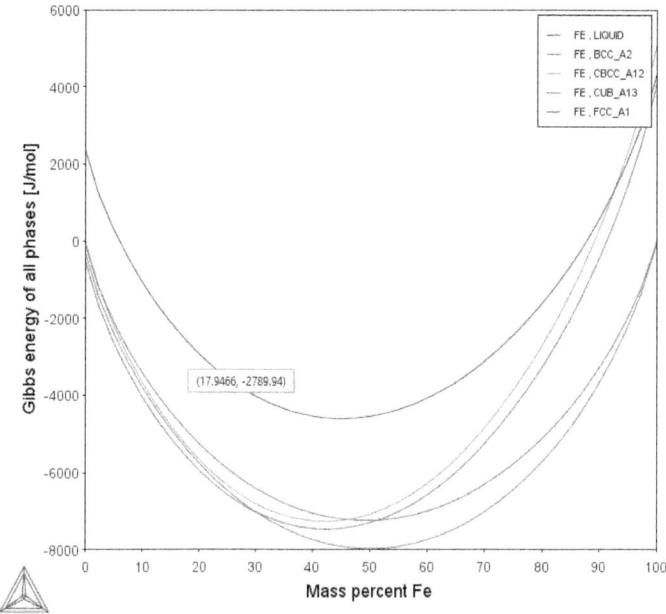

Fig:2 (a) shows gibbs free energy of Fe-Mn at 1250K (GMR). The atmospheric pressure applied 10^6 pascal for simulations

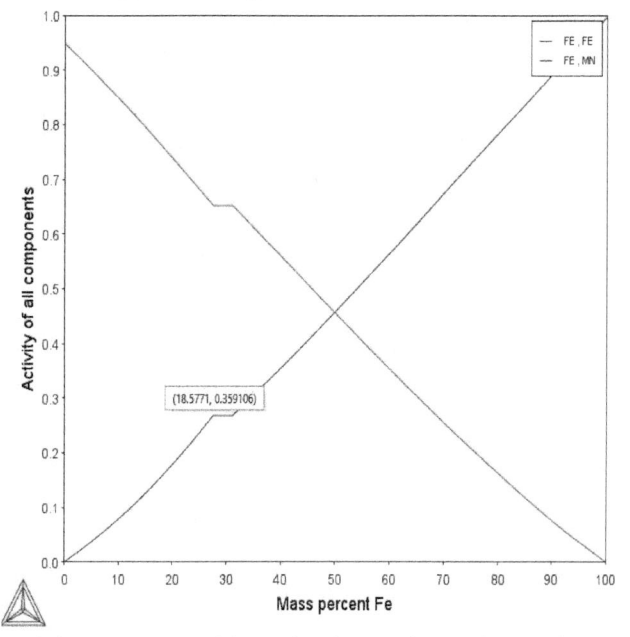

Fig:2 (b) shows Activity Fe-Mn at 1225K the thermodynamic activity vs mass fraction of stable phases is given.

Fig.2 a,b , shows corresponding curves for Gibbs free Energy and thermodynamic molar Activity of the Fe-Mn alloys system at different elevated temperature ranges (1200-1250)K respectively.The Gibbs energy curves versus different temperature show that the only phase survives at the last is FCC-A1, the negativity of this phase shows Raoult's law accordance and more stability of the Fe-Mn alloy system at this particular alloying stage. The fluctuations of other phases are also important for heterogeneity; at 1250k the most stable phase is ferrite characteristics. the proceeding table 1 showing for showing the results of the simulations in the Fe-Mn system.

The results are precise and based on database discriptions.the gibbs energy ,activity of phases vs temperature profile is assessed.

Table, 1: DATABASE THERMODYNAMIC phase CALCULATION AT TEMPERATURE: 1200K, 1225K, 1250K for the system: Fe-Mn

T°:K	Pressure:pascal	Number of moles	ACRX (activity of a component relative ratio (Mn)	Mass:gram	Total Gibbs energy:j/mol	Volume :cm3	Enthalpy:j/mol	Activity Fe. SER:stable element reference state	Activity Mn. SER: stable element reference state
1200K	1.00000*10^5	1.0000	100000*10^{-2}	5.58378*10^1	-5.72895*10^4	7.22772*10^{-6}	3.50202*10^4	9.89848*10^{-1}	1.01648*10^{-2}
--	--	--	--	--	--	--	--	9.90008*10^{-1}	1.00008*10^{-2}
--	--	--	--	--	--	--	--	3.39258*10^{-3}	1.4088*10^{-5}
--	--	--	--	--	--	--	--	-5.67338*10^4	-1.1145*10^5
FCC_A1#1	--	1.00000	--	5.5838*10^1	--	--	--	--	--
--	--	--	--	--	--	--	--	9.90000*10^{-1}	1.00000*10^{-2}
1225K	--	1.00000	1*10^{-2}	5.58378*10^1	-5.92215*10^4	7.24094*10^{-6}	3.58762*10^4	9.8984*10^{-1}	1.0164*10^{-2}
--	--	--	--	--	--	--	--	9.9000*10^{-1}	1.0000*10^{-2}
--	--	--	--	--	--	--	--	3.1547*10^{-3}	1.3321*10^{-5}
--	--	--	--	--	--	--	--	-5.8656*10^4	-1.1434*10^5
FCC_A1#1	--	1.000	--	5.5838*10^1	--	--	--	9.90000*10^{-1}	1.00000*10^{-2}
1250K	1.000008*1	1.0000	--	5.58378*10^1	-6.11710*	7.25418*10^{-6}	3.67375*10^4	9.8984*10^{-1}	1.0164*10^{-2}

	0^5				10^4				9.9000 *10-1	$1.0000*10^{-2}$
--	--	--	--	--	--	--	--	--	2.9372 *10-3	$1.2600*10^{-5}$
--	--	--	--	--	--	--	--	--	- 6.0595 *10⁴	$-1.1725*10^5$
FCC_A 1#1	--	1.0000	--	$5.5838*10^1$	--	--	--		9.9000 0*10⁻¹	$1.000008*10^{-2}$

Table 1: Shows thermodynamic fluctuations and results of Fe-Mn during alloying. at temperature of 1200K, the Gibbs energy of the alloying acquires a value of $-5.72895*10^4$ J/mol. The Enthalpy of the system reach a value of $3.50202*10^4$ J/mol corresponding to Gibbs energy .the molar activity of Fe and Mn is fluctuating a values (9.8984E-01 ,$9.90000*10^{-1}$,$3.3925*10^{-3}$,$-5.6733*10^4$,$9.9000*10^{-1}$),($1.0164*10^{-2}$,$1.0000*10^{-2}$,$1.4088*10^{-5}$,$1.00000*10^{-2}$,$-1.1145*10^5$) while the surviving phase here is FCC-A1#1 with high stability. Increasing the temperature upto 1225K, the Gibbs energy becomes $-5.92215*10^4$ J/mol with further decreasing its value, while enthalpy aquire a more positive value of $3.58762*10^4$ J/mol. The molar acrivity fluctuation at these temperature is for Fe,Mn element as($9.8984*10^{-1}$,$9.9000*10^{-1}$,$9.90000*10^{-1}$,$-5.8656*10^4$,$3.1547*10^{-3}$) ,($1.0164*10^{-2}$,$1.0000*10^{-2}$,$1.00000*10^{-2}$,$-1.1434*10^5$,$1.3321*10^{-5}$) with FCC-A1#1 phase as a stable phase for the given temperature. At 1250k of last temperature interval in our study the Total Gibbs energy reaches its highest negative value of $-6.11710*10^4$ J/mol and highest positive value of enthalpy $3.67375*10^4$ J/mol. Which shows that the system is going toward most stability even if the repulsive interaction is moreactive .The thermodynamic molar Activity of Fe,Mn elements are highest fluctuation as (9.8984E-01 ,9.9000E-01 ,2.9372E-03 ,9.90000E-01,-6.0595E+04),($1.0164*10^{-2}$,$1.0000*10^{-2}$,$1.00000*10^{-2}$,$-1.1725*10^5$,$1.2600*10^{-5}$) with highest stable phase in the phase diagram FCC-A1#1 . The activity of Fe is maximum as ($9.90000*10^{-1}$) while smaller for Mn elements. The system shows non equilibrium and stable state at highest given temperature. The Fe-Mn system is more reliable in stability and industrial needs.

4. Conclusion

The thermodynamic analysis is shown using Calphad method with pbin database of thermo-calc package, all thermodynamic optimization is of fluctuating nature, and shows compositional heterogeneity, no equilibrium is found in the said alloy system and shows metastable nature accordance. With the increasing temperature in Fe-Mn binary alloy system, the enthalpy of the system increases gradually by existing repulsion forces among alloying elements which show positive deviation from Vegard's, law, and found accordance with previous results, but the alloy still shows stability by decreasing their Gibbs energy value regularly with increasing enthalpy, which shows negative deviation from Raoult's law. The total Gibbs energy of the Fe-Mn system decreases with increasing temperature, which shows the stability of Fe-Mn system. The peak negative deviation is observed at 1250 K which shows a strong negative deviation in the Fe-Mn system and increasing the stability level means inhancing the hardness, wear resistance, corrosion resistance, and another doping characteristic of the said alloy. Ferromagnetic (FM) state of Mn-Fe alloy is the most stable one phase and occupied at the ground state. For Mn and Fe system, the measured phase equilibria show activity curve data for solid in almost ideal behavior, but somehow small positive deviation from ideality is observed. A stable phase diagram for the Fe-Mn system has been investigated in accordance with Hallowell's recommendations. The FM phase is most stable phase in that particular alloy system with occurring at the ground state. As a transition metal alloys, the magnetic properties of the Fe-Mn alloy system are very complex natured and depends on the alternation of thermal energy and curie temperatures. This favors the lower temperature. The magnetism is responsible for creating stable nature properties in that alloy system with low temperature ranges.

The enthalpy proportion with temperature shows the increase in the heat contents of the said alloy system and we observed maximum enthalpy at 1250K, which is responsible for the withstanding and surviving high temperature and high heat-absorbing capacity of Fe-Mn alloy system, activity shows throughout fluctuation from Vegard's law in a positive sense, which shows the system complex nature with rare doping and shows his greater validity for industrial variety in the system of interests and research areas.

References

1. S. Bigdeli, H. Mao Och, M. Selleby, On the third generation Calphad databases, an updated description of Mn. physica status solidi b. 252 (2015) 2199-2208.

2. K. Schwerdtfeger, Measurement of oxygen activity in iron, iron–silicon, manganese, and iron–manganese melts using solid electrolyte galvanic cells. Trans. Met. Soc. AIME . 239 (1967) 1276-1281.

3. A.T. Dinsdale, SGTE data for pure elements. Calphad. 15 (1991) 317–425.

4. W. Huang, An assessment of the Fe–Mn system. Calphad. 13 (1989) 243-252.

5. B.J. Lee, D.N. Lee, A thermodynamic study on the Mn–C and Fe–Mn systems. Calphad.13 (1989) 345-354.

6. J. Martinez, S.M. Cotes, J. Desimoni, Enthalpy change of the hcp/fcc martensitic transformation in the Fe–Mn and Fe–Mn–Si systems. J. Alloys Compounds. 479 (2009) 204–209.

7. J. Nakano, P. Jacques, Effects of the thermodynamic parameters of the hcp phase on the stacking fault energy calculations in the Fe–Mn and Fe–Mn–C systems. Calphad. 34 (2010) 167–175.

8. V.T. Witusiewicz, F. Sommer, E.J. Mittemeijer, Reevaluation of the Fe–Mn phase diagram. J. Phase Equilibria Diffusion. 25 (2004) 346–354.

9. R. Kubitz, F.H. Hayes, Enthalpies of mixing in the iron–manganese system by direct reaction Calorimetry. Mh. Chem. (1987) 31–41.

10 F. Damay, J. Sottmann, F. Lainé, L. Chaix, M. Poienar, P. Beran, E. Elkaim, F. Fauth, L. Nataf, A. Guesdon, A. Maignan, and C. Martin, Magnetic phase diagram for Fe3−xMnxBO5. Phys. Rev. B 101(2020) 094418.

11. Dejan Djurovic, Richard Dronskowski, thermodynamic assessment of Fe-Mn-c system.Calphad. 35(2011)479-491.

12. D.D. Johnson, F.J. Pinski, G.M. Stocks, Effects of chemical and magnetic disorder in Fe0.50Mn0.50. J. Appl. Phys. 63 (1988) 3490.

13. D. Birnie, E.S. Machlin, L. Kaufman, K. Taylor, Comparison of pair potential and thermochemical models of the heat of formation of BCC and FCC alloys. Calphad. 6 (1982) 93-

126.

14. A. Christou High-Pressure Phase Transition and Demagnetization in Shock Compressed Fe[Single Bond]Mn Alloys. Journal of Applied Physics. 42,11(1971) 4160 – 4170.

15 T. Gebhardt, D. Music, B. Hallstedt, M. Ekholm, I.A. Abrikosov, L. Vitos,J.M. Schneider, Ab initio lattice stability of fcc and hcp Fe–Mn random alloys, Journal of Phys: Condens. Matter, 22 (2010) 295-402.

16. A.T. Dinsdale: "Scientific Group Thermo-data Europe Database for Pure Elements," Calphad journal. 15(1991) 317-425.

17. Zhenxin Li, Yang Zhang et al, Research Progress of Fe-Based Superelastic Alloys, Crystals 12,5 (2022) 602

Printed in Great Britain
by Amazon